Herpetology of Europe and Southwest Asia:

A Checklist and Bibliography of the Orders Amphisbaenia, Sauria and Serpentes

by
KENNETH R. G. WELCH

ROBERT E. KRIEGER PUBLISHING COMPANY
MALABAR, FLORIDA
1983

Original Edition 1983

Printed and Published by
ROBERT E. KRIEGER PUBLISHING COMPANY, INC.
KRIEGER DRIVE
MALABAR, FLORIDA 32950

Printed in the United States of America

Library of Congress Cataloging in Publication Data

Welch, Kenneth R. G.
   Herpetology of Europe and southwest Asia.

   Bibliography: p.
   Includes index.
      1. Amphisbaenia—Europe.   2. Amphisbaenia—South
Asia.   3. Lizards—Europe.   4. Lizards—South Asia.
5. Snakes—Europe.   6. Snakes—South Asia.
7. Reptiles—Europe.   8. Reptiles—South Asia.
9. Reptiles—Bibliography.   I. Title
QL666.A45W44 1983   597.9'095   82-12645
ISBN 0-89874-533-0

To
my wife
Elizabeth

ACKNOWLEDGEMENTS

In preparing this checklist and preliminary bibliography I have leaned on many shoulders and would like to extend my thanks to all with particular reference to: I. S. Darevsky (U.S.S.R); D. E. Hahn (U.S.A.) for his knowledge of the Scolecophidians; G. Pasteur (France) for his knowledge of the genus *Chalcides*; N. N. Shcherbak (U.S.S.R.); and C. Edmondson and K. A. Harding for their assistance with obscure references. It must be said that all final decisions were taken by myself and as such I alone am responsible for any mistakes.

K. R. G. Welch
27 March 1982

CONTENTS

Chapter 1

CLASSIFICATION

Introduction. Over the years faunal lists for various parts of the region here covered have been presented (see Appendix 2) but no attempt has been made to list the area as a whole (Europe, excluding U.S.S.R., east through Turkey to Iran and Afghanistan south from Turkey to the Sinai Peninsula and Arabia). Such a list is readily subject to much criticism and comment and to aid such a situation I have included under each taxa, where known to myself, those references which will aid the reader to prepare descriptions and distributional maps, with information on other studies such as biology, behaviour and toxinology.

Classification. The following is intended only as a note on the classification used in the main text. The Order Amphisbaenia follows that of Gans (1967). The Order Sauria follows that of Underwood (1971) after the removal of the Amphisbaenids with two alterations: the Family Scincidae follows that of Greer (1970) with the addition of tribal names for the groups of Greer (1979); the family Lacertidae, I herein treat the subgenera of Shcherbak (1974) as full genera using the tribal name Eremiini for them as a group. The Order Serpentes follows that as used by Welch (1982).

Order Amphisbaenia
    Family Trogonophidae
    Family Amphisbaenidae
Order Sauria
  Suborder Ascalabota
    Infraorder Gekkota
      Superfamily Gekkonoidea
      Family Gekkonidae
    Infraorder Iguania
      Family Agamidae
      Family Chamaeleonidae

  Suborder Autarchoglossa
    Infraorder Scincomorpha
      Superfamily Scincoidea
      Family Scincidae
        Subfamily Scincinae
        Subfamily Lygosominae
          Tribe Sphenomorphini
          Tribe Lygosomini
      Superfamily Lacertoidea
      Family Lacertidae
        Tribe Eremiini

Infraorder Anguimorpha
Superfamily Anguioidea
Family Anguidae
Subfamily Anguinae
Superfamily Varanoidea
Family Varanidae
Order Serpentes
Suborder Scolecophidia
Family Leptotyphlopidae
Family Typhlopidae
Suborder Alethinophidia
Infraorder Henophidia
Superfamily Booidea
Family Boidae
Subfamily Boinae
Tribe Erycini
Infraorder Caenophidia
Superfamily Colubroidea
Family Colubridae
Subfamily Aparallactinae
Subfamily Atractaspidinae
Subfamily Boiginae

Subfamily Boodontinae
Subfamily Colubrinae
Subfamily Dasypeltinae
Subfamily Lycodontinae
Subfamily Lycophidinae
Subfamily Natricinae
Subfamily Philothamninae
Subfamily Psammophinae
Subfamily 'uncertain'
Superfamily Elapsoidea
Family Elapidae
Subfamily Bungarinae
Tribe Bungarini
Tribe Najini
Family Hydrophiidae
Subfamily Hydrophiinae
Tribe Hydrophiini
Superfamily Viperoidea
Family Viperidae
Subfamily Viperinae
Family Crotalidae
Subfamily Crotalinae

Chapter 2

FAMILY TROGONOPHIDAE

Order: Amphisbaenia
Family: Trogonophidae

Genus: AGAMODON Peters 1882
Species typica: *anguliceps* Peters

Agamodon arabicus Anderson 1901
    Distribution: Yemen

Genus: DIPLOMETOPON Nikolsky 1907
Species typica: *zarudnyi* Nikolsky

Diplometopon zarudnyi Nikolsky 1907
    Distribution: Western Iran, southern Iraq, Kuwait, northern Saudi
             Arabia, Trucial Oman
    Reference: Al-Nassar (1976)

Chapter 3

FAMILY AMPHISBAENIDAE

Order: Amphisbaenia
Family: Amphisbaenidae

Genus: BLANUS Wagler 1830
Species typica: *cinerea* Vandelli

Blanus cinereus cinereus (Vandelli 1797)   *Amphisbaenia*
    Distribution: Iberia (and Tangier)
    Reference: Arnold and Burton (1978); Bons (1963); Gabe and Saint
        Girons (1965) (1969); Gans (1962)

Blanus strauchi Bedriaga
    Reference: Alexander (1966)

Blanus strauchi strauchi Bedriaga 1884
    Distribution: Western Turkey and the islands of Rhodes, Kos and
        Cyprus

Blanus strauchi aporus Werner 1898
    Distribution: Turkey east from Xanthus along the Mediterranean coast
        to Lebanon and Israel, inland to eastern Iraq

Blanus strauchi bedriagae Boulenger 1884
    Distribution: Southwestern Turkey in the vicinity of Xanthus

Chapter 4

FAMILY GEKKONIDAE

Order: Sauria
Suborder: Ascalabota
Infraorder: Gekkota
Superfamily: Gekkonoidea
Family: Gekkonidae

Genus: AGAMURA Blanford 1874
Species typica: *persica* Dumeril
Reference: Schleich (1977)

Agamura cruralis Blanford 1874
  Distribution: Iran, Afghanistan (and Pakistan)

Agamura persica (Dumeril 1856)   *Gymnodactylus*
  Distribution: Iran, Afghanistan (and Pakistan)
  Reference: Anderson and Leviton (1969); Minton (1962)

Genus: ALSOPHYLAX Fitzinger 1843
Species typica: *pipiens* Pallas
Reference: Shcherbak and Golubev (1977)

Alsophylax pipiens (Pallas 1814)   *Lacerta*
  Distribution: Iran, Afghanistan (and U.S.S.R.)
  Reference: Clark and others (1969)

Genus: BUNOPUS Blanford 1874
Species typica: *tuberculatus* Blanford
Reference: Anderson (1973); Shcherbak and Golubev (1977)

Bunopus abudhabi Leviton and Anderson 1967
    Distribution: Abu Dhabi

Bunopus aspratilis Anderson 1973
    Distribution: Iran

Bunopus blanfordii Strauch 1887
    Distribution: Arabia, Israel (and Egypt)
    Reference: Haas and Werner (1969); Loveridge (1947)

Bunopus crassicauda Nikolsky 1907
    Distribution: Iran
    Reference: Minton, Anderson and Anderson (1970); Schleich (1977)

Bunopus tuberculatus Blanford 1874
    Distribution: Israel and Arabia east to Afghanistan (and Pakistan)
    Reference: Anderson (1963); Haas and Werner (1969); Schleich (1977)
    Bunopus biporus Werner: Schleich (1977)

Genus: CROSSOBAMON Boettger 1888
Species typica: *eversmanni* Wiegmann

Crossobamon eversmanni (Wiegmann 1834)   *Gymnodactylus*
    Distribution: North Iran, northwestern Afghanistan (and U.S.S.R.)

Crossobamon lumsdeni (Boulenger 1887)   *Stenodactylus*
    Distribution: Iran, Afghanistan (and Pakistan)

Crossobamon maynardi (Smith 1933)   *Stenodactylus*
    Distribution: Afghanistan (and Pakistan)

Genus: EUBLEPHARIS Gray 1827
Species typica: *hardwickii* Gray

Eublepharis angramainyu Anderson and Leviton 1966
    Distribution: southwestern Iran and Iraq

Eublepharis macularius (Blyth 1854)   *Cyrtodactylus*
  Distribution: Iraq, Iran and Afghanistan (east to India)
  Reference: Anderson (1963); Clark and others (1969); Minton (1962);
    Schleich (1977)

Eublepharis turkmenicus Darevsky 1978
  Distribution: northern Iran east of the Caspian Sea (and adjoining
    U.S.S.R.)

Genus: GYMNODACTYLUS Spix 1825
Species typica: *geckoides* Spix
Reference: Shcherbak and Golubev (1977)

Gymnodactylus agamuroides Nikolsky 1899
  Distribution: Iran
  Reference: Anderson (1963); Schleich (1977)

Gymnodactylus amictopholis (Hoofien 1967)   *Cyrtodactylus*
  Distribution: Israel

Gymnodactylus brevipes Blanford 1874
  Distribution: Iran

Gymnodactylus caspius Eichwald 1831
  Distribution: northern Iran, northern Afghanistan (and U.S.S.R.)
  Reference: Schleich (1977)

Gymnodactylus fedtschenkoi Strauch 1887
  Distribution: northern Iran, northern Afghanistan, (Pakistan and
    U.S.S.R.)
  Reference: Clark and others (1969; Schleich (1977)

Gymnodactylus heterocercus heterocercus Blanford 1874
  Distribution: Iran
  Reference: Minton, Anderson and Anderson (1970); Schleich (1977)

Gymnodactylus heterocercus mardinensis Mertens 1924
  Distribution: northeastern Turkey

Gymnodactylus kirmanensis Nikolsky 1899
  Distribution: Iran
  Reference: Schleich (1977)

Gymnodactylus longipes Nikolsky 1896
   Distribution: eastern Iran
   Reference: Orlov (1981)

Gymnodactylus scaber (Heyden 1827)   *Stendactylus*
   Distribution: Egypt east through Arabia, Iraq and Iran (to northwest
      India)
   Reference: Anderson (1963); Haas and Werner (1969); Loveridge (1947);
      Marx (1968); Schleich (1977).

Gymnodactylus turcmenicus Shcherbak 1978
   Distribution: northern Iran east of the Caspian Sea (and adjoining
      U.S.S.R.)

Gymnodactylus watsoni Murray 1892
   Distribution: Afghanistan (Pakistan and India)
   Reference: Anderson and Leviton (1969)

Genus: HEMIDACTYLUS Oken 1817
Species typica: *mabouia* Moreau de Jonnes

Hemidactylus flaviviridis Ruppell 1835
   Distribution: (northeast Africa) east through Arabia, Iraq and Iran (to
      Bengal)
   Reference: Mahendra (1935a) (1935b) (1936) (1941) (1942); Schleich
      (1977); Seshadri (1956)

Hemidactylus persicus Anderson 1872
   Distribution: Arabia, Iraq and Iran (east to northern India)
   Reference: Anderson (1963); Minton (1962); Schleich (1977)

Hemidactylus shugraensis Haas and Battersby 1959
   Distribution: southwestern Arabia

Hemidactylus turcicus turcicus (Linnaeus 1758)   *Lacerta*
   Distribution: Mediterranean coast of Europe and its islands; Turkey east
      to Iran and south to Egypt
   Reference: Baur (1979a); Jacobshagen (1937); Kostanecki (1926);
      Loveridge (1941); Pasteur and Bons (1960); Roberts and
      Schmidt-Nielsen (1966); Schleich (1977); Tansley (1959);
      Werner (1971)

Hemidactylus yerburii yerburii Anderson 1895
   Distribution: southern Arabia
   Reference: Lanza (1978); Loveridge (1941)

Genus: MEDIODACTYLUS Shcherbak and Golubev 1977
Species typica: *kotschyi* Steindachner

Mediodactylus kotschyi (Steindachner)
   Reference: Beutler and Gruber (1979); Krankenberg (1978)

Mediodactylus kotschyi kotschyi (Steindachner 1870)   *Gymnodactylus*
   Distribution: Italy, Balkan Peninsula and the Cyclades

Mediodactylus kotschyi bartoni (Stepanek 1934)   *Gymnodactylus*
   Distribution: Crete

Mediodactylus kotschyi bureschi (Stepanek 1937)   *Gymnodactylus*
   Distribution: coastal Bulgaria south to northeastern Turkey

Mediodactylus kotschyi colchicus (Nikolsky 1902)   *Gymnodactylus*
   Distribution: northeastern Turkey

Mediodactylus kotschyi fitzingeri (Stepanek 1937)   *Gymnodactylus*
   Distribution: Cyprus

Mediodactylus kotschyi kalypsae (Stepanek 1939)   *Gymnodactylus*
   Distribution: Gavdhos Island south of Crete

Mediodactylus kotschyi lycaonicus (Mertens 1952)   *Gymnodactylus*
   Distribution: southern central Turkey

Mediodactylus kotschyi oertzeni (Boettger 1888)   *Gymnodactylus*
   Distribution: southern Sporades

Mediodactylus kotschyi orientalis (Stepanek 1937)   *Gymnodactylus*
   Distribution: Turkey south to Israel

Mediodactylus kotschyi rarus (Wettstein 1952)   *Gymnodactylus*
   Distribution: islands off the southeast coast of Crete

Mediodactylus kotschyi rumelicus (L. Muller 1939)   *Gymnodactylus*
   Distribution: southern Bulgaria

Mediodactylus kotschyi saronicus (Werner 1937)   *Gymnodactylus*
   Distribution: islands of Salamis and Hydra off the Greek coast near
                Athens

Mediodactylus kotschyi steindachneri (Stepanek 1937)   *Gymnodactylus*
   Distribution: northern Turkey

Mediodactylus kotschyi stubbei (Wettstein 1952)   *Gymnodactylus*
    Distribution: Kufonisi Island off the coast of eastern Crete

Mediodactylus kotschyi syriacus (Stepanek 1937)   *Gymnodactylus*
    Distribution: eastern Turkey and Syria east to Iran

Mediodactylus kotschyi wettsteini (Stepanek 1937)   *Gymnodactylus*
    Distribution: Mikronisi Island off the coast of eastern Crete

Mediodactylus russowii (Strauch 1887)   *Gymnodactylus*
    Distribution: northern Iran, Afghanistan (and U.S.S.R.)

Mediodactylus sagittifer (Nikolsky 1899)   *Gymnodactylus*
    Distribution: Iran

Mediodactylus spinicauda (Nikolsky 1887)   *Alsophylax*
    Distribution: Iran (and U.S.S.R.)

Genus: PHYLLODACTYLUS Gray 1828
Species typica: *pulcher* Gray

Phyllodactylus elisae Werner 1895
    Distribution: Iraq and Iran
    Reference: Anderson (1963); Haas and Werner (1969); Schleich (1977)

Phyllodactylus europaeus Gene 1838
    Distribution: Corsica, Sardinia and scattered locations on the coast of
            France and western Italy; (islands off the coast of Tunisia)

    Reference: Arnold and Burton (1978); Pellegrin (1927)

Genus: PRISTURUS Ruppell 1835
Species typica: *flavipunctatus* Ruppell

Pristurus carteri carteri (Gray 1863)   *Spatalura*
    Distribution: Arabia

Pristurus carteri collaris (Steindachner 1869)   *Spatalura*
    Distribution: Hadramaut, Arabia

Pristurus carteri tuberculatus Parker 1931
    Distribution: Arabia

Pristurus crucifer (Valenciennes 1861)   *Gymnocephalus*
    Distribution: southwestern Arabia, (Ethiopia and Somalia)
    Reference: Loveridge (1947); Parker (1942); Scortecci (1935); Tornier
        (1905)

Pristurus flavipunctatus flavipunctatus Ruppell 1835
    Distribution: (northeast Africa), Israel and Arabia
    Reference: Loveridge (1947); Marx (1968)

Pristurus flavipunctatus guweirensis Haas 1951
    Distribution: Jordan

Pristurus rupestris rupestris Blanford 1874
    Distribution: (Somalia), southern Arabia and the islands of the Persian
        Gulf
    Reference: Haas and Werner (1969); Loveridge (1947); Parker (1942);
        Schleich (1977)

Pristurus rupestris iranicus Schmidt 1952
    Distribution: Iran (and Pakistan)
    Reference: Schleich (1977)

Genus: PTYODACTYLUS Goldfuss 1827
Species typica: *hasselquistii* Donndorff
Reference: Loveridge (1947); Werner (1965)

Ptyodactylus hasselquistii (Donndorff)
    Reference: Frankenberg (1974); Haas and Werner (1969); Schmidt and
        Inger (1957); Werner (1971)

Ptyodactylus hasselquistii hasselquistii (Donndorff 1798)   *Lacerta*
    Distribution: (North Africa), Israel, Jordan, Lebanon, Arabia, Syria, Iraq
        and Iran
    Reference: Tercafs (1962)

Ptyodactylus hasselquistii guttatus Heyden 1827
    Distribution: Sinai and Israel

Ptyodactylus hasselquistii puiseuxi Boutan 1893
    Distribution: Israel north to Syria
    Reference: Werner and Goldblatt (1978)

Genus: RHINOGECKO Witte 1973
Species typica: *missonnei* Witte

Rhinogecko missonnei Witte 1973
Distribution: Iran

Genus: STENODACTYLUS Fitzinger 1826
Species typica: *sthenodactylus* Lichtenstein

Stenodactylus affinis (Murray 1884)   *Ceramodactylus*
Distribution: Iran

Stenodactylus arabicus Haas 1957
Distribution: Trucial Coast, Arabia

Stenodactylus doriae (Blanford 1872)   *Ceramodactylus*
Distribution: Israel, Arabia, Iraq and Iran

Stenodactylus grandiceps Haas 1952
Distribution: Iraq

Stenodactylus khobarensis (Haas 1957)   *Pseudoceramodactylus*
Distribution: Al Khobar, Saudi Arabia

Stenodactylus leptocosymbotes Leviton and Anderson 1967
Distribution: Trucial States

Stenodactylus major (Parker 1930)   *Ceramodactylus*
Distribution: Hadramaut, Arabia
Reference: Leviton and Anderson (1967)

Stenodactylus petrii Anderson 1896
Distribution: (North Africa), Sinai and Israel
Reference: Lampe (1911); Loveridge (1947); Marx (1968); Pasteur and
        Bons (1960)

Stenodactylus pulcher Anderson 1896
Distribution: Hadramaut, Arabia

Stenodactylus slevini Haas 1957
Distribution: Saudi Arabia

Stenodactylus sthenodactylus sthenodactylus (Lichtenstein 1823)
                                                         Ascalabotes
Distribution: (North and east Africa), Israel, Arabia and Syria
Reference: Werner (1964)

Genus: TARENTOLA Gray 1825
Species typica: *mauritanica* Linnaeus
Reference: Loveridge (1947)

Tarentola annularis (Geoffroy 1827)   *Gecko*
    Distribution: Sinai (and Saharan Africa)
    Reference: Cloudsley-Thompson (1972); Grandison (1961); Hoofien
        (1962)

Tarentola mauritanica (Linnaeus 1758)   *Lacerta*
    Distribution: Iberia, Corsica, Sardinia, Balearics, Sicily, mediterranean
        coast of Europe, Southwest Asia (and North Africa), Crete
        (and the Canaries)
    Reference: Arnold and Burton (1978); Gabe (1972); Gabe and Saint
        Girons (1965) (1969); Hiller (1977); Jacobshagen (1937);
        Kostanecki (1926); Pasteur and Girot (1960); Tansley (1959);
        Wood (1938)

Genus: TERATOSCINCUS Strauch 1863
Species typica: *keyserlingii* Strauch
Reference: Anderson and Leviton (1969); Schleich (1977)

Teratoscincus bedriagai Nikolsky 1899
    Distribution: eastern Iran and Afghanistan

Teratoscincus microlepis Nikolsky 1899
    Distribution: eastern Iran, Afghanistan (and Pakistan)

Teratoscincus scincus scincus (Schlegel 1858)   *Stenodactylus*
    Distribution: Iran, Afghanistan (and U.S.S.R. east to China)

Teratoscincus scincus keyserlingii Strauch 1863
    Distribution: Iran

Genus: TRACHYDACTYLUS Haas and Battersby 1959
Species typica: *jolensis* Haas and Battersby

Trachydactylus jolensis Haas and Battersby 1959
    Distribution: Hadramaut, Arabia

Genus: TRIGONODACTYLUS Haas 1957
Species typica: *arabicus* Haas

Trigonodactylus arabicus Haas 1957
   Distribution: Arabia

Genus: TROPIOCOLOTES Peters 1880
Species typica: *tripolitanus* Peters
Reference: Anderson, S.Ç. (1961); Guibe (1966b); Leviton and Anderson
   (1972); Minton, Anderson and Anderson (1970)

Tropiocolotes helenae helenae (Nikolsky 1907)   *Microgecko*
   Distribution: Zagros Mountains, Iran
   Reference: Anderson (1963); Minton (1962)

Tropiocolotes helenae fasciatus Schmidtler and Schmidtler 1970
   Distribution: Iran

Tropiocolotes heteropholis Minton, Anderson and Anderson 1970
   Distribution: Saladehin, Iraq

Tropiocolotes latifi Leviton and Anderson 1972
   Distribution: Zagros Mountains, Iran

Tropiocolotes nattereri Steindachner 1901
   Distribution: Sinai and western Arabia
   Reference: Pasteur (1960)

Tropiocolotes persicus persicus (Nikolsky 1903)   *Alsophylax*
   Distribution: Iranian-Pakistan border

Tropiocolotes persicus bakhtiari Minton, Anderson and Anderson 1970
   Distribution: Zagros Mountains, Iran

Tropiocolotes scortecci Cherchi and Spano 1964
   Distribution: Hadramaut, Arabia

Tropiocolotes steudneri (Peters 1869)   *Gymnodactylus*
   Distribution: Israel and adjoining Jordan and Arabia (west into North
                and northeast Africa)

Chapter 5

FAMILY AGAMIDAE

Order: Sauria
Suborder: Ascalabota
Infraorder: Iguania
Family: Agamidae

Genus: AGAMA Daudin 1802
Species typica: *agama* Linnaeus

Agama agilis agilis Olivier 1804
    Distribution: Arabia north to Iraq and Iran
    Reference: Anderson (1963); Clark, Clark and Anderson (1966)

Agama agilis isolepis Boulenger 1885
    Distribution: Iran, Afghanistan (and Pakistan)
    Reference: Anderson (1966b); Haas and Werner (1969)

Agama agnetae Werner 1929
    Distribution: western Iraq
    Reference: Haas and Werner (1969)

Agama agrorensis (Stoliczka 1872)   *Stellio*
    Distribution: Afghanistan (and India)
    Reference: Anderson and Leviton (1969)

Agama blanfordi blanfordi Anderson 1966
    Distribution: Jordan, northern Arabia, southern Iraq and southwest Iran
    Reference:
    Agama persica Blanford: Anderson (1963)

Agama blanfordi fieldi Haas and Werner 1969
   Distribution: Saudi Arabia north through Jordan to Iraq
   Reference: Werner (1971)

Agama caucasia (Eichwald 1831)   *Stellio*
   Distribution: Eastern Turkey, Iraq, Iran, Afghanistan (Pakistan and U.S.S.R.)
   Reference: Ananjeva (1981); Ananjeva and Orlova (1979); Anderson and
          Leviton (1969); Arronet (1973); Clark, Clark and Anderson
          (1966); Gabaeva (1970); Luppa (1961)

Agama cyanogaster adramitana Anderson 1896
   Distribution: Yemen and Hadramaut

Agama cyanogaster yemenensis Klausewitz 1954
   Distribution: Sana, Yemen

Agama erythrogastra (Nikolsky 1896)   *Stellio*
   Distribution: northeast Iran, northern Afghanistan (and U.S.S.R.)
   Reference: Ananjeva and Orlova (1979); Anderson and Leviton (1969);
          Clark, Clark and Anderson (1966)
   Agama caucasica mucronata Guibe: Ananjeva (1981); Schleich (1977)

Agama flavimaculata (Ruppell 1835)   *Trapelus*
   Distribution: (Egypt and) western Arabia
   Reference: Pasteur and Bons (1960)

Agama himalayana himalayana (Steindachner 1869)   *Stellio*
   Distribution: northeastern Afghanistan (east to Kashmir and U.S.S.R.)
   Reference: Ananjeva (1981); Anderson and Leviton (1969)

Agama himalayana badakhshana Anderson and Leviton 1969
   Distribution: Afghanistan
   Reference: Ananjeva, Peters and Rzepakovsky (1981)

Agama jayakari Anderson 1896
   Distribution: Saudi Arabia and Oman
   Reference: Leviton and Anderson (1967)

Agama kirmanensis Nikolsky 1899
   Distribution: Iran

Agama lehmanni (Nikolsky 1896)   *Stellio*
   Distribution: Afghanistan (east to Central Asia)
   Reference: Ananjeva (1981); Anderson and Leviton (1969)

Agama melanura (Blyth 1854)   *Laudakia*
    Distribution: Iran (Pakistan and India)
    Reference: Haas and Werner (1969)

Agama microlepis (Blanford 1874)   *Stellio*
    Distribution: Iran

Agama microtympanum Werner 1895
    Distribution: Iran

Agama mutabilis Merrem 1820
    Distribution: (North Africa) Arabia, Israel, Jordan and Iraq
    Reference: Gabe and Saint Girons (1965); Jacobshagen (1937);
        Kostanecki (1926); Pasteur and Bons (1960)

Agama neumanni Tornier 1905
    Distribution: Arabia

Agama nupta nupta De Filippi 1843
    Distribution: Iraq, Iran, Afghanistan (and Pakistan)
    Reference: Anderson (1963); Clark and others (1969); Haas and Werner
        (1969)

Agama nuristanica Anderson and Leviton 1969
    Distribution: eastern Afghanistan

Agama pallida pallida Reuss 1833
    Distribution: (Egypt), Sinai and southern Israel
    Reference: Haas and Werner (1969); Werner (1971)

Agama pallida haasi Werner 1971
    Distribution: Jordan

Agama rubrigularis (Blanford 1875)   *Trapelus*
    Distribution: Iran (and Pakistan)

Agama ruderata Olivier 1807
    Distribution: Turkey, Syria, Lebanon, Jordan, northern Arabia, Iraq,
        Iran, Afghanistan, (Pakistan and U.S.S.R.)
    Reference: Ananjeva (1981); Anderson (1963); Clark and others (1969);
        Leviton and Anderson (1970b); Mertens (1952a)
    Agama megalonyx (Gunther): Wermuth (1967)

Agama savignii Dumeril and Bibron 1837
    Distribution: Israel (and Egypt)
    Reference: Marx (1968)

Agama sinaita Heyden 1827
  Distribution: (northeast Africa), Israel, Arabia, Jordan and Syria
  Reference: Werner (1971)

Agama stellio (Linnaeus)
  Reference: Baecker (1940); Daan (1967); Eyal-Giladi (1964); Jacobshagen
    (1937); Kostanecki (1926); Langerwerf (1977); Schmidt and
    Inger (1957)

Agama stellio stellio (Linnaeus 1758)   *Lacerta*
  Distribution: Greece and Corfu; Turkey south to Lebanon and east to
    Iraq
  Reference: Mertens (1952a)

Agama stellio brachydactyla Haas 1951
  Distribution: Sinai, Israel and Saudi Arabia
  Reference: Werner (1971)

Agama stellio cypriaca Daan 1967
  Distribution: Cyprus

Agama stellio picea Parker 1935
  Distribution: Jordan
  Reference: Childress (1970)

Agama tuberculata Hardwicke and Gray 1827
  Distribution: Afghanistan (east to Nepal and India)

Genus: CALOTES Cuvier 1817
Species typica: *calotes* Linnaeus

Calotes versicolor (Daudin 1802)   *Agama*
  Distribution: Iran and Afghanistan (east through Asia to Indonesia)
  Reference: Asana (1931); Baby and Reddy (1977); Bhattacharya and
    Ghose (1970); Choubey (1970); Choubey and Thapliyal
    (1966); Clark and others (1969); Goel (1976); Gouder,
    Nadkarni and Rao (1979); Indurkar and Sabnis (1976); Iyer
    (1943); Jacobshagen (1937); Kostanecki (1926); Kothari and
    Patil (1975); Mathur and Goel (1974); Minton (1962);
    Mulherkar (1962); Muthukkaruppan and others (1970);
    Pandha and Thapliyal (1964a) (1964b) (1967); Subba Rao
    (1975a) (1975b)

Genus: PHRYNOCEPHALUS Kaup 1825
Species typica: *guttatus* Gmelin
Reference: Clark and others (1969)

Phrynocephalus arabicus Anderson 1894
 Distribution: Arabia
 Reference: Leviton and Anderson (1967)
Phrynocephalus nejdensis nejdensis Haas: Schmidt and Inger (1957);
 Wermuth (1967)
Phrynocephalus nejdensis macropeltis Haas: Wermuth (1967)

Phrynocephalus clarkorum Anderson and Leviton 1967
 Distribution: Afghanistan

Phrynocephalus euptilopus Alcock and Finn 1896
 Distribution: Afghanistan (and Pakistan)

Phrynocephalus helioscopus persicus De Filippi 1863
 Distribution: Turkey east to Iran (and adjacent U.S.S.R.)
 Reference: Arronet (1973); Clark, Clark and Anderson (1966); Darevsky
 (1960); Mertens (1952a)

Phrynocephalus interscapularis interscapularis Lichtenstein 1856
 Distribution: Iran and Afghanistan (east to Central Asia)

Phrynocephalus luteoguttatus Boulenger 1887
 Distribution: Iran, Afghanistan (and Pakistan)

Phrynocephalus maculatus maculatus Andeson 1872
 Distribution: Syria, Iraq, Iran, Afghanistan (and Pakistan)

Phrynocephalus maculatus longicaudatus Haas 1957
 Distribution: northern Arabia

Phrynocephalus mystaceus galli Krassowsky 1932
 Distribution: northeastern Iran, Afghanistan (and U.S.S.R.)

Phrynocephalus ornatus Boulenger 1887
 Distribution: Iran, Afghanistan (and Pakistan)

Phrynocephalus reticulatus boettgeri Bedriaga 1907
 Distribution: northern Afghanistan (and U.S.S.R.)
 Reference: Arronet (1973)

Phrynocephalus scutellatus (Olivier 1807)   *Agama*
  Distribution: Iran, Afghanistan (and Pakistan)
  Reference: Haas and Werner (1969)

Genus: UROMASTYX Merrem 1820
Species typica: *aegyptius* Forskal
Reference: Cooper and Poole (1973); Robinson (1976)

Uromastyx aegyptius (Forskal 1775)   *Lacerta*
  Distribution: (North Africa), Sinai, Israel, Iraq and northern Arabia
  Reference: El-Toubi and Bishai (1959); Khalil and Abdel-Messeih (1954)
           (1961a); Khalil and Hussein (1962) (1963); Khalil and Yanni
           (1959) (1961); Schmidt-Nielsen and others (1963); Sokolov
           (1966); Throckmorton (1976)

Uromastyx asmussi (Strauch 1863)   *Centrotrachelus*
  Distribution: Iran, Afghanistan (and Pakistan)

Uromastyx benti (Anderson 1894)   *Aporoscelis*
  Distribution: south and southeastern Arabia

Uromastyx loricatus (Blanford 1874)   *Centrotrachelus*
  Distribution: Iraq and Iran
  Reference: Anderson (1963); Haas and Werner (1969)

Uromastyx microlepis Blanford 1874
  Distribution: Saudi Arabia, Iraq and Iran
  Reference: Haas and Werner (1969); Leviton and Anderson (1967)

Uromastyx ornatus Heyden 1827
  Distribution: (Egypt), Israel, Syria and Arabia

Uromastyx philbyi Parker 1938
  Distribution: western Arabia

Uromastyx thomasi Parker 1930
  Distribution: Hadramaut, southern Arabia.

Chapter 6

FAMILY CHAMAELEONIDAE

Order: Sauria
Suborder: Ascalabota
Infraorder: Iguania
Family: Chamaeleonidae

Genus: CHAMAELEO Laurenti 1768
Species typica: *chamaeleon* Linnaeus

Chamaeleo calyptratus Dumeril 1851
    Distribution: Yemen and southwestern Saudi Arabia

Chamaeleo chamaeleon (Linnaeus)
    Reference: Blasco (1979); Haas (1937); Ruiz (1978); Schmidt and Inger
        (1957)

Chamaeleo chamaeleon chamaeleon (Linnaeus 1758)   *Lacerta*
    Distribution: Southern Spain (and northern Africa)
    Reference: Arnold and Burton (1978); Bons and Bons (1960); Grunwald
        (1931); Haas (1937); Horn (1947); Pasteur and Bons (1960)

Chamaeleo chamaeleon calcarifer Peters 1871
    Distribution: Southern Arabian Peninsula

Chamaeleo chamaeleon musae Steindachner 1900
   Distribution: Sinai Peninsula

Chamaeleo chamaeleon orientalis Parker 1938
   Distribution: Southwestern Saudi Arabia and Yemen

Chamaeleo chamaeleon recticrista Boettger 1880
   Distribution: Israel, Lebanon and Cyprus
   Reference: Hoofien (1964); Werner (1971)

Chapter 7

FAMILY SCINCIDAE

Order: Sauria
Suborder: Autarchoglossa
Infraorder: Scincomorpha
Superfamily: Scincoidea
Family: Scincidae

Subfamily Scincinae

Genus: CHALCIDES Laurenti 1768
Species typica: *chalcides* Linnaeus
Reference: Arnold and Burton (1978); Pasteur (1981)

Chalcides bedriagai (Bosca 1880)   *Gongylus*
    Distribution: southern and eastern Iberian Peninsula

Chalcides chalcides (Linnaeus 1758)   *Lacerta*
    Distribution: Iberia, southern France, Italy, Sicily, Sardinia (and north
        west Africa)
    Reference: Schmidt and Inger (1957)

Chalcides guentheri Boulenger 1887
    Distribution: Turkey south to Israel

Chalcides levitoni Pasteur 1978
    Distribution: coastal southwest Arabia

Chalcides ocellatus (Forskal 1775)   *Lacerta*
    Distribution: Sardinia, Sicily, Malta, Crete, Cyprus, Greece east through
            Turkey and Southwest Asia to Pakistan and south to North
            Africa
    Reference: Badir (1958) (1959) (1968a) (1968b); Badir and Hussein
            (1965); Khalil (1951); Parker (1942); Richter (1933); Tercafs
            (1962); Trost (1953); Vogel (1964)

Chalcides pistaciae Valverde 1967
    Distribution: Portugal and central Spain

Genus: EUMECES Wiegmann 1834
Species typica: *pavimentatus* Geoffroy
Reference: Taylor (1935)

Eumeces schneiderii (Daudin)
    Reference: Eiselt (1940); Mertens (1920) (1946); Smirnov (1979)

Eumeces schneiderii schneiderii (Daudin 1802)   *Scincus*
    Distribution: Cyprus; Syria south to Arabia (west to Algeria)

Eumeces schneiderii blythianus (Anderson 1871)   *Mabouia*
    Distribution: Afghanistan (?) (and adjoining India)

Eumeces schneiderii pavimentatus (Geoffroy 1827)   *Scincus*
    Distribution: Syria south to Egypt

Eumeces schneiderii princeps (Eichwald 1839)   *Euprepes*
    Distribution: Jordan east to Iran
    Reference: Werner (1971)

Eumeces schneiderii variegatus Schmidt 1939
    Distribution: Iran
    Reference: Schleich (1977)

Eumeces schneiderii zarudnyi Nikolsky 1899
    Distribution: Iran (east into Pakistan)
    Reference: Haas and Werner (1969); Schleich (1977)

Eumeces taeniolatus (Blyth 1854)   *Eurylepis*
    Distribution: Iran and Afghanistan (east into India)
    Reference: Minton (1962)

Genus: OPHIOMORUS Dumeril and Bibron 1839
Species typica: *punctatissimus* Bibron and Bory St. Vincent
Reference: Anderson and Leviton (1966)

Ophiomorus blanfordi Boulenger 1887
Distribution: coastal sand dunes of southern Iran (and Pakistan)

Ophiomorus brevipes (Blanford 1874)   *Zygnopsis*
Distribution: sandy regions of eastern Iran
Reference: Anderson (1963)

Ophiomorus latastii Boulenger 1887
Distribution: Lebanon, Israel and Jordan

Ophiomorus nuchalis Nilson and Andren 1978
Distribution: Kavir Desert, Iran

Ophiomorus persicus (Steindachner 1867)   *Hemipodion*
Distribution: western slopes of Zagros Mountains, Iran

Ophiomorus punctatissimus (Bibron and Bory St. Vincent 1833)   *Anguis*
Distribution: southwestern Turkey and Greece
Reference: Clark and Clark (1970)

Ophiomorus streeti Anderson and Leviton 1966
Distribution: southeastern Iran

Ophiomorus tridactylus (Blyth 1855)   *Sphenocephalus*
Distribution: eastern Iran, southern Afghanistan (and Pakistan)
Reference: Clark and others (1969); Minton (1962); Pough (1971);
Rathor (1970)

Genus: SCINCUS Laurenti 1768
Species typica: *scincus* Linnaeus
Reference: Arnold and Leviton (1977)

Scincus hemprichii Wiegmann 1837
Distribution: extreme southwestern Arabian Peninsula

Scincus mitranus Anderson 1871
Distribution: south and east Arabia

Scincus scincus (Linnaeus)
>   Reference: Badir (1958) (1959); Badir and Hussein (1965); El-Toubi
>   (1938); Gabe and Saint Girons (1965); Richter (1933)

Scincus scincus scincus (Linnaeus 1758)   *Lacerta*
>   Distribution: (Egypt), Sinai and Israel

Scincus scincus conirostris Blanford 1881
>   Distribution: south and eastern Arabia, Iraq and southwestern Iran
>   Reference: Anderson (1963)

Scincus scincus meccensis Wiegmann 1837
>   Distribution: northwest Arabia and southern Jordan

Genus: SPHENOPS Wagler 1830
Species typica: *sepsoides* Audouin

Sphenops sepsoides (Audouin 1827)   *Scincus*
>   Distribution: Jordan, Israel and Sinai (west to West Africa)
>   Reference: Khalil and Hussein (1962) (1963); Marx (1968); Pasteur and
>   Bons (1960); Werner (1968)

Subfamily Lygosominae
Tribe Sphenomorphini

Genus: ABLEPHARUS Lichtenstein 1823
Species typica: *pannonicus* Lichtenstein
Reference: Fuhn (1969)

Ablepharus bivittatus bivittatus (Menetries 1832)   *Scincus*
>   Distribution: eastern Turkey, northwest and north Iran (and U.S.S.R.)

Ablepharus bivittatus lindbergi Wettstein 1960
>   Distribution: western Afghanistan

Ablepharus kitaibelii kitaibelii Bibron and Bory St. Vincent 1833
>   Distribution: Greece, Aegean Islands, Rhodes, Cyprus, Turkey, Syria,
>   Israel, Jordan, Lebanon, Iraq and Sinai

Ablepharus kitaibelii fabichi Stepanek 1938
   Distribution: islands of Mikronisi, Karpathos, Kasos and Armathia off
      the east coast of Crete

Ablepharus kitaibelii fitzingeri Mertens 1952
   Distribution: Hungary and Czechoslovakia
   Reference: Vogel (1964)

Ablepharus kitaibelii stepaneki Fuhn 1969
   Distribution: Romania and Bulgaria

Ablepharus pannonicus pannonicus Lichtenstein 1823
   Distribution: Jordan, Syria, Iraq, Iran, Afghanistan (and U.S.S.R.)
   Reference: Leviton and Anderson (1963)

Ablepharus pannonicus grayanus (Stoliczka 1872)   *Blepharosteres*
   Distribution: eastern Iran, Afghanistan, (Pakistan and U.S.S.R.)

Genus: SCINCELLA Mittleman 1950
Species typica: *lateralis* Say

Scincella himalayana (Gunther 1864)   *Eumeces*
   Distribution: eastern Afghanistan (east to Nepal)
   Reference: Duda and Koul (1977)

Subfamily Lygosominae
Tribe Lygosomini

Genus: MABUYA Fitzinger 1826
Species typica: *mabouya* Lacepede

Mabuya aurata (Linnaeus 1758)   *Lacerta*
   Distribution: Cyprus; Turkey east to Iran (and adjoining U.S.S.R.) south
            through Iraq, and Syria to the Persian Gulf of Arabia
   Reference: Anderson (1963); Haas and Werner (1969); Mertens (1952a);
            Schleich (1977)

Mabuya brevicollis (Wiegmann 1837)   *Euprepis*
   Distribution: southwest Arabia (and eastern Africa)

Mabuya dissimilis (Hallowell 1860)   *Euprepes*
    Distribution: Afghanistan (east into India)
    Reference: Anderson and Leviton (1969); Clark and others (1969)

Mabuya tessellata Anderson 1895
    Distribution: southwest Arabia

Mabuya vittata (Olivier 1804)   *Scincus*
    Distribution: (North Africa) and the Sinai north to Iraq
    Reference: Haas and Werner (1969); Werner (1971)

Chapter 8

FAMILY LACERTIDAE

Order: Sauria
Suborder: Autarchoglossa
Infraorder: Scincomorpha
Superfamily: Lacertoidea
Family: Lacertidae

Genus: ACANTHODACTYLUS Wiegmann 1834
Species typica: *boskianus* Daudin
Reference: Boulenger (1921)

Acanthodactylus boskianus (Daudin)
> Reference: Duvdevani (1972); Duvdevani and Borut (1974); Kostanecki (1926); Schmidt and Inger (1957)

Acanthodactylus boskianus boskianus (Daudin 1802)   *Lacerta*
> Distribution: Sinai (and coastal Egypt)

Acanthodactylus boskianus asper (Audouin 1829)   *Lacerta*
> Distribution: (North Africa east to Ethiopia) north to Iraq; Saudi Arabia
> Reference: Angel and Lhote (1938); Haas and Werner (1969); Werner (1971)

Acanthodactylus boskianus euphraticus Boulenger 1921
> Distribution: Iraq

Acanthodactylus boskianus schreiberi Boulenger 1878
> Distribution: Cyprus

Acanthodactylus boskianus syriacus Boettger 1879
  Distribution: Syria

Acanthodactylus cantoris cantoris Gunther 1864
  Distribution: southern Afghanistan (Pakistan and India)

Acanthodactylus cantoris arabicus Boulenger 1918
  Distribution: Sinai south through Arabia

Acanthodactylus cantoris blanfordi Boulenger 1918
  Distribution: southeastern Iran (and southwest Pakistan)
  Reference: Anderson (1963); Haas and Werner (1969); Schleich (1977)

Acanthodactylus cantoris schmidti Haas 1957
  Distribution: Saudi Arabian coast of the Persian Gulf, southwest Iran
    and Iraq
  Reference: Anderson (1963); Leviton and Anderson (1967); Schleich
    (1977)

Acanthodactylus erythrurus erythrurus (Schinz 1833)   *Lacerta*
  Distribution: Iberia
  Reference: Arnold and Burton (1978); Kostanecki (1926)

Acanthodactylus fraseri Boulenger 1918
  Distribution: Iraq

Acanthodactylus gongrorhynchatus Leviton and Anderson 1967
  Distribution: Saudi Arabia, Abu Dhabi and Trucial Coast

Acanthodactylus grandis Boulenger 1909
  Distribution: Syria and Jordan
  Reference: Haas and Werner (1969); Werner (1971)

Acanthodactylus haasi Leviton and Anderson 1967
  Distribution: Saudi Arabia

Acanthodactylus micropholis Blanford 1874
  Distribution: Iran (and Pakistan)
  Reference: Haas and Werner (1969); Schleich (1977)

Acanthodactylus pardalis (Lichtenstein 1823)   *Lacerta*
  Distribution: Israel and Sinai (west to Libya)
  Reference: Bons (1960); Busack (1975); Duvdevani (1972); Duvdevani
    and Borut (1974)

Acanthodactylus robustus Werner 1929
    Distribution: Syria
    Reference: Haas and Werner (1969)

Acanthodactylus scutellatus (Audouin)
    Reference: Duvdevani (1972); Duvdevani and Borut (1974)

Acanthodactylus scutellatus scutellatus (Audouin 1829)   *Lacerta*
    Distribution: Israel and Sinai (west to Algeria)
    Reference: Bons and Girot (1963)

Acanthodactylus scutellatus hardyi Haas 1957
    Distribution: Saudi Arabia north to Iraq
    Reference: Haas and Werner (1969)

Acanthodactylus tristrami tristrami (Gunther 1864)   *Zootoca*
    Distribution: Lebanon, Jordan and Syria
    Reference: Werner (1971)

Acanthodactylus tristrami iracensis Schmidt 1939
    Distribution: Iraq
    Reference: Haas and Werner (1969)

Acanthodactylus tristrami orientalis Angel 1936
    Distribution: northeast Syria and Iraq
    Reference: Haas (1952); Haas and Werner (1969); Riney (1953); Schmidt
            (1939)

Genus: ALGYROIDES Bibron and Bory St. Vincent 1833
Species typica: *moreoticus* Bibron and Bory St. Vincent
Reference: Arnold (1973); Jacobshagen (1937); Klemmer (1960); Kostanecki
            (1926); Mertens and Wermuth (1960)

Algyroides fitzingeri (Wiegmann 1834)   *Notopholis*
    Distribution: Corsica and Sardinia

Algyroides marchi Valverde 1958
    Distribution: Sierra de Cazorla, southeast Spain
    Reference: Eikhorst and others (1979)

Algyroides moreoticus Bibron and Bory St. Vincent 1833
    Distribution: southern Greece and the Ionian Islands
    Reference: Clark and Clark (1970)

Algyroides nigropunctatus (Dumeril and Bibron 1839)   *Lacerta*
Distribution: Istria south to northwest Greece; Ionian Islands

Genus: LACERTA Linnaeus 1758
Species typica: *agilis* Linnaeus
Reference: Arnold (1973); Mertens and Wermuth (1960)

Lacerta agilis Linnaeus
    Reference: Avery (1976); Baecker (1940); Baranoff and Valetzky (1975);
        Beebee (1978); Borchwardt (1977); Corbett and Tamarind
        (1979); Eggert (1935); Gabaeva (1970); Glandt (1976) (1977);
        Hett (1924); Jacobshagen (1937); Koch (1904); Kostanecki
        (1926); Langerwerf (1980); Luppa (1961); Marx and Kayser
        (1949); Regamey (1936); Saint Girons (1976); Simms (1976);
        Sjongren (1945); Spellerberg (1974) (1976); Spitz (1971);
        Stolk (1953); Swiezawska (1949); Vogel (1964)

Lacerta agilis agilis Linnaeus 1758
    Distribution: France, Belgium, Holland, Denmark, England, Sweden,
        Germany, Poland, Austria, Switzerland, Hungary,
        Czechoslovakia, northwest Yugoslavia, western Romania
        (and U.S.S.R.)
    Reference: Borcea (1979); Schulz (1972)

Lacerta agilis bosnica Schreiber 1912
    Distribution: southern Yugoslavia and Bulgaria

Lacerta agilis chersonensis Andrzejowski 1832
    Distribution: south and east Romania (and U.S.S.R.)
    Reference: Vladescu (1965b); Vladescu and Motelica (1965)

Lacerta armeniaca Mehely 1909
    Distribution: extreme eastern Turkey (and U.S.S.R.)
    Reference: Arronet (1973); Darevsky (1966); Darevsky and Danielyan
        (1968); Darevsky, Kupriyanova and Bakradze (1978);
        Langerwerf (1980a); Uzzell and Darevsky (1975)

Lacerta bedriagae bedriagae Camerano 1885
    Distribution: Corsica

Lacerta bedriagae paessleri Mertens 1927
    Distribution: Limbara Mountains, Sardinia

Lacerta bedriagae sardoa Peracca 1903
Distribution: Gennargentu Mountains, Sardinia

Lacerta bithynica Mehely 1909
Distribution: Turkey

Lacerta brandti De Fillipi 1863
Distribution: northwest Iran (and U.S.S.R.)
Reference: Clark, Clark and Anderson (1966); Lantz and Cyren (1939);
        Schleich (1977)

Lacerta cappadocica cappadocica Werner 1902
Distribution: central Turkey east to northwest Iran
Reference: Eiselt (1979)

Lacerta cappadocica urmiana (Lantz and Suchow 1934)   *Apathya*
Distribution: Iran
Reference: Schleich (1977)

Lacerta cappadocica wolteri (Bird 1936)   *Apathya*
Distribution: Turkish-Syrian border

Lacerta chlorogaster Boulenger 1908
Distribution: Iran (and U.S.S.R.)

Lacerta clarkorum Darevsky and Vedmederja 1977
Distribution: northeast Turkey (and U.S.S.R.)
Reference: Darevsky and Lukina (1977)

Lacerta cyanura Arnold 1972
Distribution: southeastern Arabia

Lacerta danfordii (Gunther)
Reference: Wettstein (1967)

Lacerta danfordii danfordii (Gunther 1876)   *Zootoca*
Distribution: southeast Turkey east to Iran and south to Jordan
Reference: Schleich (1977)

Lacerta danfordii anatolica Werner 1900
Distribution: west and northwest Turkey
Reference: Langerwerf (1980a)

Lacerta danfordii kulzeri Muller and Wettstein 1933
Distribution: southwest Turkey

Lacerta danfordii oertzeni Werner 1904
   Distribution: Ikaria Island south of Izmir, Turkey

Lacerta danfordii pelasgiana Mertens 1959
   Distribution: Rhodes, near Turkey

Lacerta danfordii pentanisiensis Wettstein 1964
   Distribution: Pentanisos Island, east of Rhodes

Lacerta defillipi (Camerano 1877)   *Podarcis*
   Distribution: Turkey and northern Iran
   Reference: Arronet (1973); Clark, Clark and Anderson (1966); Darevsky
         (1966); Schleich (1977)

Lacerta derjugini Nikolsky 1898
   Distribution: northern Turkey (and U.S.S.R.)
   Reference: Bischoff (1974a); Polozhikhina (1965); Uzzell and Darevsky
         (1973)

Lacerta fraasii Lehrs 1910
   Distribution: Lebanon
   Reference: Peters (1962a)

Lacerta graeca Bedriaga 1886
   Distribution: Taygetos Mountains, Greece

Lacerta horvathi Mehely 1904
   Distribution: northwest Yugoslavia and adjacent Italy

Lacerta jayakari Boulenger 1887
   Distribution: southeastern Arabia

Lacerta laevis Gray 1838
   Distribution: southeast Turkey south to Israel; Cyprus
   Reference: Langerwerf (1980a)

Lacerta lepida lepida Daudin 1802
   Distribution: Iberia, southern France and northwest Italy
   Reference: Allen (1977); Beguin (1902); Lopez and Bons (1981);
         Schmidt and Inger (1957); Weber (1977)

Lacerta mehelyi Lantz and Cyren 1936
   Distribution: Turkey (and U.S.S.R.)

Lacerta monticola monticola Boulenger 1905
Distribution: Sierra Estrella, Portugal

Lacerta monticola bonnali Lantz 1927
Distribution: region of Lake de Bigorre in the Pyrenees

Lacerta monticola cantabrica Mertens 1929
Distribution: Cantabrian Mountains, Spain

Lacerta monticola cyreni Muller and Hellmich 1937
Distribution: Sierra de Guadarama and de Gredos, Spain
Reference: Langerwerf (1980a)

Lacerta mosorensis Kolombatovic 1886
Distribution: southwest Yugoslavia

Lacerta oxycephala Dumeril and Bibron 1839
Distribution: southwest Yugoslavia

Lacerta parva Boulenger 1887
Distribution: Turkey (and U.S.S.R.)
Reference: Lantz and Cyren (1939); Mertens (1952a); Peters (1962a)

Lacerta parvula parvula Lantz and Cyren 1913
Distribution: northeastern Turkey
Reference: Darevsky and Lukina (1977); Uzzell and Darevsky (1975)

Lacerta parvula adjarica Darevsky and Eiselt 1980
Distribution: extreme coastal northeast Turkey (and U.S.S.R.)

Lacerta perspicillata perspicillata Dumeril and Bibron 1839
Distribution: Menorca (introduced) (and northwest Africa)

Lacerta praticola pontica Lantz and Cyren 1919
Distribution: Romania, eastern Yugoslavia, Bulgaria, European Turkey
    (and U.S.S.R.)
Reference: Bischoff (1976b); Langerwerf (1980a)

Lacerta princeps princeps Blanford 1874
Distribution: eastern Turkey east to southwest Iran
Reference: Eiselt (1968) (1969); Mertens (1952a) (1952b); Schleich
    (1977)

Lacerta princeps kurdistanica Suchov 1936
Distribution: Zagros Mountains, Iran

Lacerta raddei raddei Boettger 1892
   Distribution: northwest Iran (and U.S.S.R.)
   Reference: Uzzell and Darevsky (1973b) (1975)

Lacerta raddei nairensis Darevsky 1967
   Distribution: northeast Turkey (and U.S.S.R.)
   Reference: Danielyan (1965); Darevsky (1967)

Lacerta rudis Bedriaga
   Reference: Bischoff (1974c); Darevsky and Eiselt (1980); Darevsky and
      Lukina (1977); Langerwerf (1980a)

Lacerta rudis rudis Bedriaga 1886
   Distribution: northeast Turkey (and U.S.S.R.)

Lacerta rudis bischoffi Bohme and Budak 1977
   Distribution: coastal northeast Turkey (and U.S.S.R.)

Lacerta rudis tristis Lantz and Cyren 1933
   Distribution: northeast Turkey (and U.S.S.R.)

Lacerta schreiberi Bedriaga 1878
   Distribution: northwest, west and central Iberia

Lacerta strigata strigata Eichwald 1831
   Distribution: Iran, Iraq, eastern Turkey (north into U.S.S.R.)
   Reference: Bischoff (1976a); Haas and Werner (1969); Ivanov and
      Fedorova (1970); Langerwerf (1980a) (1980b); Peters
      (1962b)

Lacerta trilineata Bedriaga
   Reference: Langerwerf (1980a); Peters (1962b) (1964b)

Lacerta trilineata trilineata Bedriaga 1886
   Distribution: Balkan Peninsula north to Istria; Ionian Islands and the
      Cyclades (excluding Milos, Kimolos and Sifnos)

Lacerta trilineata dobrogica Fuhn and Mertens 1959
   Distribution: eastern Romania

Lacerta trilineata hansschweizeri Muller 1935
   Distribution: Cyclades: Milos, Kimolos and Sifnos

Lacerta trilineata media Lantz and Cyren 1920
Distribution: eastern Turkey, Iran and Iraq (north into U.S.S.R.)
Reference: Bischoff (1974b) (1975b); Clark, Clark and Anderson (1966);
Fuhn (1956); Haas and Werner (1969); Luppa (1961);
Schleich (1977)

Lacerta trilineata polylepidota Wettstein 1952
Distribution: Crete

Lacerta unisexualis Darevsky 1966
Distribution: northeast Turkey (and U.S.S.R.)
Reference: Darevsky and Danielyan (1968) (1979); Uzzell and Darevsky
(1975)

Lacerta uzzelli Darevsky and Danielyan 1977
Distribution: Eastern Turkey
Reference: Darevsky and Lukina (1977)

Lacerta valentini Boettger
Reference: Darevsky (1967); Darevsky and Lukina (1977)

Lacerta valentini valentini Boettger 1889
Distribution: northeast Turkey (and U.S.S.R.)
Reference: Arronet (1973); Danielyan (1965); Uzzell and Darevsky
(1975)

Lacerta valentini lantzicyreni Darevsky and Eiselt 1967
Distribution: eastern Turkey

Lacerta viridis (Laurenti)
Reference: Appleyard (1978) (1979b); Bailey (1969); Beguin (1902);
Blanchard (1890); Ducker and Rensch (1973); Foxon,
Griffith and Price (1956); Gerzeli and Piceis Polver (1970);
Jacobs (1979); Jacobshagen (1937); Kostanecki (1926);
Langerwerf (1980a); Perschmann (1956); Peters (1962b);
Plehn (1911); Quattrini (1953); Raynaud and Adrian (1976)
(1977); Saint Girons (1976) (1977); Schnabel (1954);
Spellerberg (1974); Spitz (1971); Weber (1957)

Lacerta viridis viridis (Laurenti 1768)   *Seps*
Distribution: northern Spain, France, Switzerland, northern Italy,
Germany, Austria, Czechoslovakia, Poland, Balkan
Peninsula and Euboa (east to U.S.S.R.)

Lacerta viridis chloronota Rafinesque 1810
  Distribution: Calabaria and Sicily

Lacerta viridis citrovittata Werner 1938
  Distribution: Tinos, Cyclades

Lacerta viridis fejervaryi Vasvary 1926
  Distribution: Campania and Puglia, Italy

Lacerta viridis meridionalis Cyren 1933
  Distribution: Romania, eastern Bulgaria and Turkey east to Iran

Lacerta viridis woosnami Boulenger 1917
  Distribution: Iranian coast of the Caspian Sea (and U.S.S.R.)

Lacerta vivipara Jacquin 1787
  Distribution: North and West Europe south to northern Spain and east
        through Central Europe (U.S.S.R. and northern Mongolia)
  Reference: Andersen (1971); Armstrong (1950); Avery (1962) (1966)
        (1970) (1971) (1973) (1975a) (1975b) (1976); Avery and
        McArdle (1973); Avery, Shewry and Stobart (1974); Baur
        (1979b); Bea (1978b); Dufaure (1961) (1966) (1968) (1970)
        (1971); Dufaure and Hubert (1965); Eggert (1935); Glandt
        (1976) (1977); Hubert (1965) (1967) (1968b) (1968c) (1970)
        (1971a) (1971b) (1972) (1976); Hubert and Xavier (1979);
        Luppa (1961); Moffat and Bellairs (1972); Pearson and
        Tamarind (1973); Petzold (1978); Quattrini (1953); Schmidt
        and Inger (1957); Sheppard and Bellairs (1972); Spellerberg
        (1976); Weigmann (1932)

Genus: LATASTIA Bedriaga 1884
Species typica: *boscai* Bedriaga
Reference: Boulenger (1921); Parker (1942)

Latastia longicaudata longicaudata (Reuss 1834)   *Lacerta*
  Distribution: Sinai (north and northeast Africa)

Latastia longicaudata andersonii Boulenger 1921
  Distribution: southwest Arabia

Genus: OPHISOPS Menetries 1832
Species typica: *elegans* Menetries

Ophisops elegans Menetries 1832
  Distribution: southern Bulgaria and northeast Greece east through Asia
      Minor to India and south to Egypt (Libya and Sudan);
      Cyprus
  Reference: Anderson (1963); Arronet (1973); Bischoff (1974d);
      Boulenger (1921)

Ophisops jerdoni Blyth 1853
  Distribution: eastern Afghanistan (and Pakistan)

Genus: PHILOCHORTUS Matschie 1893
Species typica: *neumanni* Matschie

Philochortus neumanni Matschie .1893
  Distribution: southwest Arabia

Genus: PODARCIS Wagler 1830
Species typica: *muralis* Laurenti
Reference: Arnold (1973)

Podarcis bocagei (Seoane 1884)   *Lacerta*
  Distribution: northwest Spain and northern Portugal

Podarcis erhardii (Bedriaga)
  Reference: Langerwerf (1980a); Walter (1967)

Podarcis erhardii erhardii (Bedriaga 1882)   *Lacerta*
  Distribution: Cyclades: Seriphos and Siphnos

Podarcis erhardii amorgensis (Werner 1933)   *Lacerta*
  Distribution: Cyclades: Amorgos Island

Podarcis erhardii biinsulicola (Wettstein 1937)   *Lacerta*
  Distribution: Makri Island near Anafi, Cyclades

Podarcis erhardii buchholzi (Wettstein 1956)   *Lacerta*
  Distribution: Ktenia Island east of Naxos, Cyclades

Podarcis erhardii cretensis (Wettstein 1952)   *Lacerta*
  Distribution: northwest Crete

Podarcis erhardii elaphonisii (Wettstein 1952)   *Lacerta*
  Distribution: Elafonisos Island off the coast of Greece

Podarcis erhardii gaigeae (Werner 1930)   *Lacerta*
    Distribution: Skiros Island, Northern Sporades

Podarcis erhardii kinarensis (Wettstein 1937)   *Lacerta*
    Distribution: Kinaros Island northeast of Amorgos, Cyclades

Podarcis erhardii leukaorii (Wettstein 1952)   *Lacerta*
    Distribution: western Crete

Podarcis erhardii levithensis (Wettstein 1937)   *Lacerta*
    Distribution: Levitha Island, Cyclades

Podarcis erhardii livadiaca (Werner 1902)   *Lacerta*
    Distribution: Central Greece and Peloponnise; Euboea

Podarcis erhardii makariaisi (Wettstein 1956)   *Lacerta*
    Distribution: northern Makariais Islands, Cyclades

Podarcis erhardii megalophthenae (Wettstein 1937)   *Lacerta*
    Distribution: Megalo Petali, Cyclades

Podarcis erhardii mykonensis (Werner 1933)   *Lacerta*
    Distribution: Mikonos, Tinos, Andros and Dhilos islands, Cyclades

Podarcis erhardii naxensis (Werner 1899)   *Lacerta*
    Distribution: Cyclades: Naxos, Paros, Iraklia, Ios, Sikinos, Folegandros,
        Thira and Thirasia islands

Podarcis erhardii obscura (Wettstein 1952)   *Lacerta*
    Distribution: Islands of Mikronisi and Gaidhouronisi off the south coast
        of Crete
    Reference:
    Lacerta erhardii werneriana Wettstein: Mertens and Wermuth (1960)

Podarcis erhardii ophidusae (Wettstein 1937)   *Lacerta*
    Distribution: Cyclades: Ofidhousa Island

Podarcis erhardii pachiae (Wettstein 1937)   *Lacerta*
    Distribution: Cyclades: Pachia Island south of Anafi

Podarcis erhardii phytiusae (Wettstein 1937)   *Lacerta*
    Distribution: Cyclades: Skhoinousa and Phytiusa islands

Podarcis erhardii psathurensis (Cyren 1941)   *Lacerta*
    Distribution: Northern Sporades: Psathoura Island

Podarcis erhardii punctigularis (Wettstein 1952)   *Lacerta*
    Distribution: Prassonisi Cliffs, southwest Crete

Podarcis erhardii rechingeri (Wettstein 1952)   *Lacerta*
    Distribution: Dragonada Island, northeast of Crete

Podarcis erhardii riveti (Chabanaud 1919)   *Lacerta*
    Distribution: southern Bulgaria, southern Yugoslavia, Albania and
        northern Greece

Podarcis erhardii ruthveni (Werner 1930)   *Lacerta*
    Distribution: Northern Sporades: Kyra Panagia and Giura

Podarcis erhardii schiebeli (Wettstein 1952)   *Lacerta*
    Distribution: Dia Island off the north coast of Crete

Podarcis erhardii scopelensis (Cyren 1941)   *Lacerta*
    Distribution: Northern Sporades: Skopelos and Iliodromia

Podarcis erhardii subobscura (Wettstein 1937)   *Lacerta*
    Distribution: Tria Nisia Island south of Sirna, Aegean Sea

Podarcis erhardii syrinae (Wettstein 1937)   *Lacerta*
    Distribution: Sirna Island, Aegean Sea

Podarcis erhardii thermiensis (Werner 1935)   *Lacerta*
    Distribution: Cyclades: Kithnos Island

Podarcis erhardii thessalica (Cyren 1938)   *Lacerta*
    Distribution: eastern Central Greece

Podarcis erhardii zafranae (Wettstein 1937)   *Lacerta*
    Distribution: Zafora Island, south of Astipalaia, Aegean Sea

Podarcis filfolensis filfolensis (Bedriaga 1876)   *Lacerta*
    Distribution: Filfera (= Filfola) Island near Malta

Podarcis filfolensis generalensis (Gulia 1914)   *Lacerta*
    Distribution: General's (= Fungus) Island, west of Gozo

Podarcis filfolensis kieselbachi (Fejervary 1924)   *Lacerta*
    Distribution: San Paul Island near Malta

Podarcis filfolensis laurentiimuelleri (Fejervary 1924)   *Lacerta*
    Distribution: Lampione and Linosa Islands west of Malta

Podarcis filfolensis maltensis Mertens 1921
  Distribution: Malta and Gozo

Podarcis hispanica (Steindachner)
  Reference: Gabe (1971)(1972); Gabe and Saint Girons (1972); Guillaume
    (1976); Jacobshagen (1937); Kostanecki (1926)

Podarcis hispanica hispanica (Steindachner 1870)   *Lacerta*
  Distribution: east and central Spain and Mediterranean France

Podarcis hispanica atrata (Bosca 1916)   *Lacerta*
  Distribution: Columbretes off the Mediterranean coast of Spain

Podarcis hispanica vaucheri (Boulenger 1905)   *Lacerta*
  Distribution: southern Spain (and North West Africa)

Podarcis lilfordi (Gunther)
  Reference: Salvador (1980)

Podarcis lilfordi lilfordi (Gunther 1874)   *Zootoca*
  Distribution: Balearics: Ayre Island southeast of Menorca

Podarcis lilfordi balearica (Bedriaga 1879)   *Lacerta*
  Distribution: Balearics: Rey, Addaya, Sargantana and Robello Islands off
    the coast of Menorca

Podarcis lilfordi brauni (Muller 1927)   *Lacerta*
  Distribution: Balearics: Colon Island near Menorca

Podarcis lilfordi conejerae (Muller 1927)   *Lacerta*
  Distribution: Balearics: Conejera Island off the south coast of Mallorca

Podarcis lilfordi fahrae (Muller 1927)   *Lacerta*
  Distribution: Balearics: Horadada and Pobre Islands south of Mallorca

Podarcis lilfordi fenni (Eisentraut 1928)   *Lacerta*
  Distribution: Balearics: Porros Island off the north coast of Menorca

Podarcis lilfordi gigliolii (Bedriaga 1879)   *Lacerta*
  Distribution: Balearics: Dragonera Island southwest of Mallorca

Podarcis lilfordi hartmanni (Wettstein 1937)   *Lacerta*
  Distribution: Balearics: Isla Malgrats southwest of Mallorca

Podarcis lilfordi jordansi (Muller 1927)   *Lacerta*
Distribution: Balearics: islands of Guardia, Moltona and Islota de Frailes
south

Podarcis lilfordi kuligae (Muller 1927)   *Lacerta*
Distribution: Cabrera Island south of Mallorca

Podarcis lilfordi planae (Muller 1927)   *Lacerta*
Distribution: Balearics: Plana Island south of Mallorca

Podarcis lilfordi rodriquezi (Muller 1927)   *Lacerta*
Distribution: Balearics: Ratas Island, Bay of Mahon, Menorca

Podarcis lilfordi toronis (Hartmann 1953)   *Lacerta*
Distribution: Balearcis: Toro Island, Mallorca

Podarcis melisellensis (Braun)
Reference: Clover (1979); Thorpe (1980b)

Podarcis melisellensis melisellensis (Braun 1877)   *Lacerta*
Distribution: Melisello Island, Yugoslavia

Podarcis melisellensis aeoli (Radovanovic 1959)   *Lacerta*
Distribution: Mali Opuh Island, Yugoslavia

Podarcis melisellensis bokicae (Radovanovic 1956)   *Lacerta*
Distribution: Vrtlac Island near Molat, Yugoslavia

Podarcis melisellensis curzolensis (Taddei 1950)   *Lacerta*
Distribution: Korcula Island, Yugoslavia

Podarcis melisellensis digenea (Wettstein 1926)   *Lacerta*
Distribution: Svetak Island west of Vis, Yugoslavia

Podarcis melisellensis fiumana (Werner 1891)   *Lacerta*
Distribution: Istria south along the Yugoslav coast to Albania

Podarcis melisellensis galvagnii (Werner 1908)   *Lacerta*
Distribution: Kamik Island near Svetak, Yugoslavia

Podarcis melisellensis gigantea (Radovanovic 1956)   *Lacerta*
Distribution: islands opposite Dubrovnik, Yugoslavia

Podarcis melisellensis gigas (Wettstein 1926)   *Lacerta*
Distribution: Mali Parsanj Island, Yugoslavia

Podarcis melisellensis gracilis (Radovanovic 1951)   *Lacerta*
    Distribution: Ciovo Island, near Split, Yugoslavia
    Reference:
    Lacerta melisellensis traguriana Radovanovic: Mertens and Wermuth
                    (1960)

Podarcis melisellensis jidulae (Radovanovic 1959)   *Lacerta*
    Distribution: Jidula Island, Yugoslavia

Podarcis melisellensis kammereri (Wettstein 1926)   *Lacerta*
    Distribution: Mali Barjak Island, Yugoslavia

Podarcis melisellensis kornatica (Radovanovic 1959)   *Lacerta*
    Distribution: Kornati Islands, Yugoslavia

Podarcis melisellensis lissana (Werner 1891)   *Lacerta*
    Distribution: Lissa and Lagosta Islands, Yugoslavia

Podarcis melisellensis mikavicae (Radovanovic 1959)   *Lacerta*
    Distribution: Mikavica Island, Yugoslavia

Podarcis melisellensis plutonis (Radovanovic 1959)   *Lacerta*
    Distribution: Jerolim Island near Hvar, Yugoslavia

Podarcis melisellensis pomoensis (Wettstein 1926)   *Lacerta*
    Distribution: Pomo Island, Yugoslavia

Podarcis melisellensis thetidis (Radovanovic 1959)   *Lacerta*
    Distribution: Veli Puh Island, Yugoslavia

Podarcis milensis milensis (Bedriaga 1882)   *Lacerta*
    Distribution: Cyclades: Milos, Kimolos and Agios Eustathios

Podarcis milensis gerakuniae (Muller 1938)   *Lacerta*
    Distribution: Cyclades: Falkonera northwest of Andimilos

Podarcis milensis schweizeri (Mertens 1934)   *Lacerta*
    Distribution: Cyclades: Andimilos

Podarcis muralis (Laurenti)
    Reference: Appleyard (1979a); Avery (1976) (1978); Bachmann (1979);
                Beguin (1902); Boag (1973); Eggert (1935); Gabe and Saint
                Girons (1965) (1969); Guillaume (1976); Jacobshagen (1937);

Joly and Saint Girons (1975); Koch (1904); Kostanecki (1926); Licht, Hoyer and Oordt (1969); Maher (1961); Marx and Kayser (1949); Sjongren (1945); Steward (1965); Stolk (1953); Weber (1957)

Podarcis muralis muralis (Laurenti 1768)   *Seps*
   Distribution: Holland and Germany south to the Pyrenees and east to
      Hungary, Romania, Greece and Turkey
   Reference: Froesch-Franzon (1974); Vogel (1964)

Podarcis muralis albanica (Bolkay 1919)   *Lacerta*
   Distribution: south Yugoslavia, Albania, western Greece and
      Peloponneses
   Reference: Schmidt and Inger (1957)

Podarcis muralis beccarii (Lanza 1958)   *Lacerta*
   Distribution: Italy: Islet of Port Ercole and Mount Argentario

Podarcis muralis breviceps (Boulenger 1905)   *Lacerta*
   Distribution: Italy: Calabaria

Podarcis muralis brueggemanni (Bedriaga 1879)   *Lacerta*
   Distribution: Italy: Tuscany
   Reference: Horn (1976)

Podarcis muralis calbia (Blanchard 1891)   *Lacerta*
   Distribution: France: Pointe du Raz

Podarcis muralis colosii (Taddei 1949)   *Lacerta*
   Distribution: Elba

Podarcis muralis insulanica (Bedriaga 1881)   *Lacerta*
   Distribution: Pianosa south of Elba

Podarcis muralis maculiventris (Werner 1891)   *Lacerta*
   Distribution: northeast Italy south along the Yugoslav coast to include
      the Velebit Mountains

Podarcis muralis marcuccii (Lanza 1956)   *Lacerta*
   Distribution: Argentarola Isle off the Tuscany coast

Podarcis muralis merremia (Risso 1826)   *Lacerta*
   Distribution: French Mediterranean coast

Podarcis muralis muellerlorenzi (Taddei 1949)   *Lacerta*
    Distribution: La Scuola Isle south of Elba

Podarcis muralis nigriventris Bonaparte 1836
    Distribution: Italy: vicinity of Rome

Podarcis muralis oyensis (Blanchard 1891)   *Lacerta*
    Distribution: France: Ile d'Yeu

Podarcis muralis paulinii (Taddei 1953)   *Lacerta*
    Distribution: Italy: Monte Argentario, Tuscany

Podarcis muralis rasquinetii (Bedriaga 1878)   *Lacerta*
    Distribution: northwest Spain: La Deva Island

Podarcis muralis tinettoi (Taddei 1949)   *Lacerta*
    Distribution: Italy: Tinetto Island near La Spezia

Podarcis muralis vinciguerrai (Mertens 1932)   *Lacerta*
    Distribution: Italy: Gorgona Island, Ligurian Sea

Podarcis peloponnesiaca (Bibron and Bory St. Vincent 1833)   *Lacerta*
    Distribution: southern Greece
    Reference: Buchholz (1960)

Podarcis pityusensis (Bosca)
    Reference: Wittig (1976)

Podarcis pityusensis pityusensis (Bosca 1883)   *Lacerta*
    Distribution: Balearics: Ibiza and Mallorca (introduced)

Podarcis pityusensis affinis (Muller 1927)   *Lacerta*
    Distribution: Balearics: Malvin Pequeno Isle east of Ibiza

Podarcis pityusensis algae (Wettstein 1937)   *Lacerta*
    Distribution: Alga (= Pouet) Island, north of Formentera

Podarcis pityusensis calaesaladae (Muller 1928)   *Lacerta*
    Distribution: Balearics: Cala Salada west of Ibiza

Podarcis pityusensis caldesiana (Muller 1928)   *Lacerta*
    Distribution: Balearics: Caldes Isle north of Ibiza

Podarcis pityusensis canensis (Eisentraut 1928)   *Lacerta*
    Distribution: Balearics: Cana Island east of Ibiza

Podarcis pityusensis caragolensis (Buchholz 1954)   *Lacerta*
     Distribution: Balearics: Caragole (= Negretta) Island

Podarcis pityusensis carlkochi (Mertens and Muller 1940)   *Lacerta*
     Distribution: Cunillera Island and Del Bosque west of Ibiza

Podarcis pityusensis characae (Buchholz 1954)   *Lacerta*
     Distribution: Balearics: Isla Characa, Ibiza

Podarcis pityusensis espalmadoris (Muller 1928)   *Lacerta*
     Distribution: Balearics: Isla Espalmador

Podarcis pityusensis formenterae (Eisentraut 1928)   *Lacerta*
     Distribution: Balearics: Formentera

Podarcis pityusensis frailensis (Eisentraut 1928)   *Lacerta*
     Distribution: Balearics: Isla del Fraile, west of Ibiza

Podarcis pityusensis gastabiensis (Eisentraut 1928)   *Lacerta*
     Distribution: Balearics: islands of Gastabi, Negra, Ahorcados and
               Espardell between Ibiza and Formentera

Podarcis pityusensis grossae (Muller 1929)   *Lacerta*
     Distribution: Balearics: Isla Grossa (= St. Eulalia) east of Ibiza

Podarcis pityusensis grueni (Muller 1928)   *Lacerta*
     Distribution: Balearics: Isla dos Trocados near Espalmador

Podarcis pityusensis hedwigkamerae (Muller 1927)   *Lacerta*
     Distribution: Balearics: Margalida (= Margarita) Isle off Ibiza

Podarcis pityusensis hortae (Buchholz 1954)   *Lacerta*
     Distribution: Balearics: Isla Horta northwest of Isla de Tagomago

Podarcis pityusensis kameriana (Mertens 1927)   *Lacerta*
     Distribution: Balearics: Isla Esparto west of Ibiza
     Reference: Ouboter (1976)

Podarcis pityusensis maluquerorum Mertens 1921
     Distribution: Balearics: islands of Bleda Plana, Bleda Bosque, Bleda
               Gorra and Escui de Vermey near Ibiza
     Reference: Vogel (1964)

Podarcis pityusensis miguelensis (Eisentraut 1928)   *Lacerta*
     Distribution: Balearics: Isla del Bosque de San Miguel north of Ibiza

Podarcis pityusensis muradae (Eisentraut 1928)   *Lacerta*
   Distribution: Balearics: Isla Murada west of Ibiza

Podarcis pityusensis puercosensis (Buchholz 1954)   *Lacerta*
   Distribution: Balearics: Isla Puercos

Podarcis pityusensis purroigensis (Buchholz 1954)   *Lacerta*
   Distribution: Balearics: Isleta de Purroige, Ibiza

Podarcis pityusensis ratae (Eisentraut 1928)   *Lacerta*
   Distribution: Balearics: Isla Ratas southwest of the port of Ibiza

Podarcis pityusensis redonae (Eisentraut 1928)   *Lacerta*
   Distribution: Balearics: Insel Redonda east of Ibiza

Podarcis pityusensis sabinae (Buchholz 1954)   *Lacerta*
   Distribution: Balearics: Isla Sabina

Podarcis pityusensis schreitmuelleri (Muller 1927)   *Lacerta*
   Distribution: Balearics: Malvin Grande Island, Ibiza

Podarcis pityusensis subformenterae (Buchholz 1954)   *Lacerta*
   Distribution: Balearics: Coneja de Formentera

Podarcis pityusensis tagomagensis (Muller 1927)   *Lacerta*
   Distribution: Balearics: Isla de Tagomago

Podarcis pityusensis torretensis (Buchholz 1954)   *Lacerta*
   Distribution: Balearics: Isla Torretas

Podarcis pityusensis vedrae (Muller 1927)   *Lacerta*
   Distribution: Balearics: Isla del Vedra southwest of Ibiza
   Reference: Ouboter (1976)

Podarcis pityusensis zenonis (Muller 1928)   *Lacerta*
   Distribution: Balearics: Escui de Esparto west of Ibiza

Podarcis sicula (Rafinesque)
   Reference: Avery (1976) (1978); Botte (1973a) (1973b) (1974); Botte,
      Angelini and Picariello (1978); Botte and Delrio (1965);
      Clover (1979); Licht, Hoyer and Oordt (1969); Quattrini
      (1952a) (1952b) (1954); Taddei (1972); Thorpe (1980b);
      Spellerberg (1976); Spellerberg and Smith (1975)

Podarcis sicula sicula (Rafinesque 1810)   *Lacerta*
 Distribution: Italy south of Rome, Sicily, Aeolian Islands and Pantelleria
 Reference: Filosa (1973)

Podarcis sicula adriatica (Werner 1902)   *Lacerta*
 Distribution: Pelagoso Picciola Island, Adriatic Sea

Podarcis sicula alvearioi (Mertens 1955)   *Lacerta*
 Distribution: Aeolian Islands: Faraglione Pollara northwest of Salina

Podarcis sicula astorgae (Mertens 1937)   *Lacerta*
 Distribution: Astorga Island west of Istria

Podarcis sicula bagnolensis (Mertens 1937)   *Lacerta*
 Distribution: Bagnole Island west of Istria

Podarcis sicula bolei (Brelih 1961)   *Lacerta*
 Distribution: Istrian island of Tovarez

Podarcis sicula campestris De Betta 1857
 Distribution: Italy north of Rome south along the Yugoslav coast to
              Split; Elba, Corsica and Monte Christo
 Reference: Fischer (1970); Marin and Sabbadin (1959); Schulz (1972)

Podarcis sicula cazzae (Schreiber 1912)   *Lacerta*
 Distribution: Adriatic island of Cazza

Podarcis sicula cerbolensis (Taddei 1949)   *Lacerta*
 Distribution: Cerboli Island near Elba

Podarcis sicula cettii (Cara 1872)   *Lacerta*
 Distribution: Sardinia

Podarcis sicula ciclopica (Taddei 1949)   *Lacerta*
 Distribution: Isola de Trezza (= Ciclopi) east of Sicily

Podarcis sicula coerulea (Eimer 1872)   *Lacerta*
 Distribution: Faraglioni Rocks near Capri

Podarcis sicula dupinici (Radovanovic 1956)   *Lacerta*
 Distribution: islands of Mali Dupinic and Veliki Dupinic off shore from
              Sibenik, Yugoslavia

Podarcis sicula flavigula (Mertens 1937)   *Lacerta*
 Distribution: San Giovanni Faro Island west of Istria

Podarcis sicula gallensis (Eimer 1881)   *Lacerta*
    Distribution: Italy: Galli Island, Gulf of Salerno

Podarcis sicula hadzii (Brelih 1961)   *Lacerta*
    Distribution: Porer on the west Istrian coast

Podarcis sicula hieroglyphica (Berthold 1842)   *Lacerta*
    Distribution: Istanbul and the islands of the Marmara Sea

Podarcis sicula insularum (Mertens 1937)   *Lacerta*
    Distribution: islands of La Longa, Galopon, San Giovanni in Pelago,
        Rivera, San Marco, Zumpin Piccolo and Gronghera off the
        west coast of Istria

Podarcis sicula kolombatovici (Karaman 1928)   *Lacerta*
    Distribution: small islands near Trogir, coastal Yugoslavia

Podarcis sicula laganjensis (Radovanovic 1956)   *Lacerta*
    Distribution: Adriatic islands of Mali Laganj and Veliki Laganj

Podarcis sicula latastei (Bedriaga 1879)   *Lacerta*
    Distribution: Ponza Island in the Gulf of Gaeta, Italy

Podarcis sicula liscabiancae (Mertens 1952)   *Lacerta*
    Distribution: Aeolian Islands; Lisca Bianca

Podarcis sicula major (Mertens 1916)   *Lacerta*
    Distribution: Paestum and Giungano, Gulf of Salerno, Italy
    Reference:
    Lacerta sicula mertensi Wettstein: Mertens and Wermuth (1960)

Podarcis sicula medemi (Mertens 1942)   *Lacerta*
    Distribution: Isola Bella near Taormina, Sicily

Podarcis sicula mediofasciata (Radovanovic 1959)   *Lacerta*
    Distribution: islands of Duzac and Mala Sestrica west of Istria

Podarcis sicula monaconensis (Eimer 1881)   *Lacerta*
    Distribution: Monacone Rocks near Capri

Podarcis sicula nikolici (Brelih 1961)   *Lacerta*
    Distribution: Gusti island off the west coast of Istria

Podarcis sicula pasquinii (Lanza 1966)   *Lacerta*
    Distribution: Scoglio Cappello islet south of Palmarola, Ponziane Islands

Podarcis sicula patrizii (Lanza 1952)   *Lacerta*
    Distribution: Zannone Island in the Gulf of Gaeta, Italy

Podarcis sicula pelagosae (Bedriaga 1886)   *Lacerta*
    Distribution: Pelagosa Grande in the Adriatic

Podarcis sicula pirosoensis (Mertens 1937)   *Lacerta*
    Distribution: Piroso Grande in the Adriatic

Podarcis sicula polenci (Brelih 1961)   *Lacerta*
    Distribution: Calbula on the west Istrian coast

Podarcis sicula premudana (Radovanovic 1959)   *Lacerta*
    Distribution: Premuda and adjoining islands off the Yugoslav coast

Podarcis sicula premudensis (Radovanovic 1959)   *Lacerta*
    Distribution: Lutrosnjak Island northwest of Premuda

Podarcis sicula pretneri (Brelih 1961)   *Lacerta*
    Distribution: Gustinja and Pisulj on the west Istrian coast

Podarcis sicula radovanovici (Brelih 1961)   *Lacerta*
    Distribution: Orata on the west Istrian coast

Podarcis sicula raffonei (Mertens 1952)   *Lacerta*
    Distribution: Aeolian Islands: Strombolicchio near Stromboli

Podarcis sicula ragusae (Wettstein 1931)   *Lacerta*
    Distribution: vicinity of Dubrovnik, Yugoslavia

Podarcis sicula roberti (Taddei 1949)   *Lacerta*
    Distribution: Formiche di Grosseto, Italy

Podarcis sicula salfii (Lanza 1954)   *Lacerta*
    Distribution: Vivaro di Nerano, south coast of Sorrento Peninsula, Italy

Podarcis sicula samogradi (Radovanovic 1956)   *Lacerta*
    Distribution: islands of Samograd and Vrtlic, Dalmatia, Yugoslavia

Podarcis sicula sanctinicolai (Taddei 1949)   *Lacerta*
    Distribution: San Nicola near Tremiti in the Italian Adriatic

Podarcis sicula sanctistephani (Mertens 1926)   *Lacerta*
    Distribution: San Stefano near Ventotene, Gulf of Gaeta, Italy

Podarcis sicula trischittai (Mertens 1952)   *Lacerta*
  Distribution: Aeolian Islands: Bottaro east of Panaria

Podarcis sicula tyrrhenica (Mertens 1932)   *Lacerta*
  Distribution: Giglio Islands in the Toscano Archipelago; Capraia north
       of Elba

Podarcis sicula ventotenensis (Taddei 1949)   *Lacerta*
  Distribution: Ventotene, Gulf of Gaeta, Italy

Podarcis sicula vesseljuchi (Radovanovic 1959)   *Lacerta*
  Distribution: Vesseljuh Island, Yugoslav Adriatic

Podarcis sicula zeii (Brelih 1961)   *Lacerta*
  Distribution: Kal on the west Istrian coast

Podarcis taurica taurica (Pallas 1814)   *Lacerta*
  Distribution: Hungary, Romania, Bulgaria, east and south Yugoslavia,
       Albania, northern Greece and European Turkey (east into
       U.S.S.R.)
  Reference: Cruce (1977); Cruce and Leonte (1973); Mertens (1952a);
       Vogel (1964)

Podarcis taurica ionica (Lehrs 1902)   *Lacerta*
  Distribution: Corfu, Ionian Islands and southern Greece

Podarcis taurica thasopulae (Kattinger 1942)   *Lacerta*
  Distribution: Thasopoula Island near Thasos in the Aegean Sea

Podarcis tiliguerta tiliguerta (Gmelin 1789)   *Lacerta*
  Distribution: Corsica and Sardinia

Podarcis tiliguerta ranzii (Lanza 1966)   *Lacerta*
  Distribution: Molarotto islet east of Molara, northeast Sardinia

Podarcis tiliguerta toro (Mertens 1932)   *Lacerta*
  Distribution: Toro Island southwest of Sardinia

Podarcis wagleriana wagleriana Gistel 1868
  Distribution: Sicily and the Egadi Islands of Favignana and Levanzo

Podarcis wagleriana antoninoi (Mertens 1955)   *Lacerta*
  Distribution: Aeolian island of Vulcano

Podarcis wagleriana marettimensis (Klemmer 1956)   *Lacerta*
    Distribution: the Egadi Island of Marettimo

Genus: PSAMMODROMUS Fitzinger 1826
Species typica: *hispanicus* Fitzinger
Reference: Arnold (1973); Boulenger (1921)

Psammodromus algirus algirus (Linnaeus 1758)   *Lacerta*
    Distribution: Iberia, mediterranean France (and northwest Africa)
    Reference: Schmidt and Inger (1957)

Psammodromus hispanicus hispanicus Fitzinger 1826
    Distribution: western and southern Iberia

Psammodromus hispanicus edwarsianus (Duges 1829)   *Lacerta*
    Distribution: eastern Iberia and mediterranean France

Tribe Eremiini

Genus: EREMIAS Wiegmann 1834
Species typica: *velox* Pallas
Reference: Shcherbak (1971) (1974)

Eremias persica Blanford 1874
    Distribution: eastern Iran, Afghanistan (Pakistan and U.S.S.R.)
    Reference: Bogdanov and Vashetko (1972); Clark, Clark and Anderson
        (1966); Clark and others (1969)

Eremias regeli Bedriaga 1905
    Distribution: Afghanistan (and U.S.S.R.)
    Reference: Clark and others (1969)

Eremias strauchi Kessler
    Reference: Arronet (1973); Vashetko (1969)

Eremias strauchi strauchi Kessler 1878
    Distribution: northwest Iran, eastern Turkey (and U.S.S.R.)
    Reference: Mertens (1952a)

Eremias strauchi kopetdaghica Shcherbak 1972
    Distribution: northeast Iran (and U.S.S.R.)

Eremias velox velox (Pallas 1771)   *Lacerta*
    Distribution: Iran (and U.S.S.R.)
    Reference: Ananjeva (1977); Arronet (1973); Vashetko (1971)

Genus: MESALINA Gray 1838
Species typica: *rubropunctata* Lichtenstein
Reference: Shcherbak (1974)

Mesalina adramitana (Boulenger 1917)   *Eremias*
    Distribution: Hadramut, Arabia

Mesalina brevirostris brevirostris Blanford 1874
    Distribution: Sinai north to Syria and east (to northwest India)
    Reference: Haas and Werner (1969); Hoofien (1957)

Mesalina brevirostris fieldi (Haas and Werner 1969)   *Eremias*
    Distribution: southwest Iran

Mesalina brevirostris microlepis (Angel 1936)   *Eremias*
    Distribution: Jordan north to western Syria

Mesalina guttulata (Lichtenstein)
    Reference: Haas and Werner (1969)

Mesalina guttulata guttulata (Lichtenstein 1823)   *Lacerta*
    Distribution: (North Africa) Sinai east to Iraq and Iran
    Reference: Blanc (1979); Werner (1971)

Mesalina guttulata watsonana (Stoliczka 1872)   *Eremias*
    Distribution: Iran, Afghanistan (east into India)
    Reference: Anderson (1963); Clark, Clark and Anderson (1966); Clark
        and others (1969); Leviton and Anderson (1963)

Mesalina mucronata (Blanford 1870)   *Acanthodactylus*
    Distribution: Sinai (south to Somalia)

Mesalina olivieri schmidti (Haas 1951)   *Eremias*
    Distribution: Sinai and Israel

Mesalina rubropunctata (Lichtenstein 1823)   *Lacerta*
    Distribution: Sinai (west to Algeria)

Genus: OMMATEREMIAS Lantz 1928
Species typica: *arguta* Pallas
Reference: Shcherbak (1974)

Ommateremias arguta (Pallas)
  Reference: Ivanov and Fedorova (1973)

Ommateremias arguta deserti (Gmelin 1789)   *Lacerta*
  Distribution: eastern Turkey and northwest Iran (north into U.S.S.R.);
              eastern Romania (east into U.S.S.R.)
  Reference: Tertyshnikov (1970)

Ommateremias arguta transcaucasica (Darevsky 1953)   *Eremias*
  Distribution: eastern Turkey, northwest Iran (and U.S.S.R.)

Ommateremias aria (Anderson and Leviton 1967)   *Eremias*
  Distribution: eastern Afghanistan

Ommateremias nigrocellata (Nikolsky 1896)   *Eremias*
  Distribution: northeast Iran, Afghanistan (and U.S.S.R.)
  Reference: Anderson and Leviton (1969)

Genus: RHABDEREMIAS Lantz 1928
Species typica: *scripta* Strauch
Reference: Shcherbak (1974)

Rhabderemias andersoni (Darevsky and Shcherbak 1978)   *Eremias*
  Distribution: Iran

Rhabderemias fasciata (Blanford 1874)   *Eremias*
  Distribution: eastern Iran, southern Afghanistan (and Pakistan)

Rhabderemias lineolata (Nikolsky 1896)   *Scapteira*
  Distribution: eastern Iran, Afghanistan (and U.S.S.R.)
  Reference: Ananjeva (1977)

Rhabderemias pleskei (Bedriaga 1907)   *Eremias*
  Distribution: northwest Iran, eastern Turkey (and U.S.S.R.)
  Reference: Clark, Clark and Anderson (1966)

Rhabderemias scripta (Strauch 1867)   *Podarces*
  Distribution: Iran (?), Afghanistan (Pakistan and U.S.S.R.)
  Reference: Ananjeva (1977)

Genus: SCAPTEIRA Wiegmann 1834
Species typica: *grammica* Lichtenstein
Reference: Shcherbak (1974)

Scapteira acutirostris Boulenger 1887
Distribution: Afghanistan (and Pakistan)

Scapteira aporosceles Alcock and Finn 1896
Distribution: Afghanistan (and Pakistan)

Scapteira grammica (Lichtenstein 1823)   *Lacerta*
Distribution: Iran, Afghanistan (east through U.S.S.R. to China)
Reference: Ananjeva (1977); Tzellarius (1977)

Chapter 9

FAMILY ANGUIDAE

Order: Sauria
Suborder: Autarchoglossa
Infraorder: Anguimorpha
Superfamily: Anguioidea
Family: Anguidae
Subfamily: Anguinae

Genus: ANGUIS Linnaeus 1758
Species typica: *fragilis* Linnaeus

Anguis fragilis Linnaeus
    Reference: Ali (1950); Baecker (1940); Baker (1942); Bannister (1968);
        Beguin (1902); Dalcq (1920) (1921); Dufaure (1968); Duguy
        (1963); Gabe (1972); Gabe and Saint Girons (1969) (1972);
        Gregory (1980); Greschik (1917); Hubert (1968a) (1971c);
        Jacobshagen (1937); Jeuniaux (1963a); Kostanecki (1926);
        Pernkopf and Lehner (1937); Raynaud (1959) (1960) (1961)
        (1962); Saint Girons and Saint Girons (1956); Sjongren
        (1945); Spellerberg (1976); Trost (1953)

Anguis fragilis fragilis Linnaeus 1758
    Distribution: Europe (and northwestern Africa)

Anguis fragilis colchicus (Nordmann 1840)   *Otophis*
    Distribution: Romania and Bulgaria (east into U.S.S.R.) and Turkey east
        to Iran

Anguis fragilis peloponnesiacus Stepanek 1937
   Distribution: Peloponnesia, Greece
   Reference: Clark and Clark (1970)

Genus: OPHISAURUS Daudin 1803
Species typica: *ventralis* Linnaeus

Ophisaurus apodus (Pallas 1775)   *Lacerta*
   Distribution: Istria south through the Balkans and east through Turkey
           and Syria to Iran (to include adjoining U.S.S.R.) south from
           Turkey to Jordan
   Reference: Clark and Clark (1970); Elkan (1976); Schmidt and Inger
           (1957)

Chapter 10

FAMILY VARANIDAE

Order: Sauria
Suborder: Autarchoglossa
Infraorder: Anguimorpha
Superfamily: Varanoidea
Family: Varanidae

Genus: VARANUS Merrem 1820
Species typica: *varius* Shaw

Varanus bengalensis bengalensis (Daudin 1802)   *Tupinambis*
    Distribution: Southeastern Iran and Afghanistan (east to Burma)
    Reference: Auffenberg (1979); Bhattacharya (1921); Clark and others
           (1969); Deraniyagala (1958); Gabe (1972); Gabe and Saint
           Girons (1972); Kostanecki (1926); Minton (1962); Thapar
           (1921); Vogel (1964)

Varanus griseus (Daudin)
    Reference: Gabe and Saint Girons (1965); Kostanecki (1926); Pernkopf
           and Lehner (1937)

Varanus griseus griseus (Daudin 1803)   *Tupinambis*
    Distribution: Syria, Lebanon, Israel, Jordan, Iraq and the Arabian
           Peninsula (west across northern Africa)

Varanus griseus caspius (Eichwald 1831)   *Psammosaurus*
    Distribution: Iran, Afghanistan, (Pakistan and U.S.S.R.)
    Reference: Anderson (1963); Minton (1962); Vogel (1964)

Chapter 11

FAMILY LEPTOTYPHLOPIDAE

Order: Serpentes
Suborder: Scolecophidia
Family: Leptotyphlopidae

Genus: LEPTOTYPHLOPS Fitzinger 1843
Species typica: *nigricans* Schlegel
Reference: Hahn (1978a) (1978b)

Leptotyphlops blanfordi blanfordi (Boulenger 1890)   *Glauconia*
    Distribution: Afghanistan, (India and Pakistan)

Leptotyphlops blanfordi nursi (Boulenger 1896)   *Glauconia*
    Distribution: Southern Arabian Peninsula

Leptotyphlops buri (Boulenger 1905)   *Glauconia*
    Distribution: Southwestern Arabian Peninsula

Leptotyphlops hamulirostris (Nikolsky 1907)   *Glauconia*
    Distribution: Iran

Leptotyphlops macrorhynchus macrorhynchus (Jan 1861)   *Stenostoma*
    Distribution: (West and northern Africa), southwestern Asia (east to
                India and) north to Turkey
    Reference:
    Leptotyphlops phillipsi Barbour: Hahn (1978a); Werner and Drook
                (1967)

60

Chapter 12

FAMILY TYPHLOPIDAE

Order: Serpentes
Suborder: Scolecophidia
Family: Typhlopidae

Genus: RAMPHOTYPHLOPS Fitzinger 1843
Species typica: *multilineatus* Schlegel

Ramphotyphlops braminus (Daudin 1803)   *Eryx*
    Distribution: Arabian Peninsula

Genus: TYPHLOPS Oppel 1811
Species typica: *lumbricalis* Linnaeus

Typhlops simoni (Boettger 1879)   *Onychocephalus*
    Distribution: Israel and Syria

Typhlops vermicularis Merrem 1820
    Distribution: Greece, Turkey, Syria, Israel, Lebanon, (Egypt), Iraq,
                Iran, Afghanistan (and Pakistan); Albania, Yugoslavia and
                Bulgaria
    Reference: Gabe and Saint Girons (1965); Kostanecki (1926); Steward
                (1971); Vogel (1964)

Typhlops wilsoni Wall 1908
    Distribution: Southwestern Iran

61

Chapter 13

FAMILY BOIDAE

Order: Serpentes
Suborder: Alethinophidia
Infraorder: Henophidia
Superfamily: Booidea
Family: Boidae
Subfamily: Boinae
Tribe: Erycini

Genus: ERYX Daudin 1803
Species typica: *turcica* Olivier
Reference: Rage (1972)

Eryx colubrinus (Linnaeus 1758)   *Anguis*
    Distribution: Yemen (and northeastern Africa south to Tanzania)
    Reference: Scortecci (1932)

Eryx elegans (Gray 1849)   *Cusoria*
    Distribution: Northeastern Iran, Afghanistan, (adjoining U.S.S.R. and
                India)
    Reference: Anderson and Leviton (1969)

Eryx jaculus jaculus (Linnaeus 1758)   *Anguis*
    Distribution: Syria, Iraq, Iran, Israel, Jordan and northern Saudi Arabia
                (west across northern Africa to Morocco)
    Reference: Haas (1930) (1931a)

Eryx jaculus familiaris Eichwald 1831
    Distribution: Northwestern Iran and eastern Turkey (north into U.S.S.R.)
    Reference: Steward (1971)

Eryx jaculus turcicus (Olivier 1801)   *Boa*
  Distribution: Romania, Yugoslavia, Albania, Corfu, Greece, the
            Cyclades and western Turkey
  Reference: Stemmler (1958); Steward (1971)

Eryx jayakari Boulenger 1888
  Distribution: Eastern Arabian Peninsula
  Reference: Corkill and Cochrane (1966)

Eryx johnii (Russell)
  Reference: Gabe and Saint Girons (1969); Minton (1962)

Eryx johnii johnii (Russell 1801)   *Boa*
  Distribution: Afghanistan (east into India)

Eryx johnii persicus Nikolsky 1907
  Distribution: Iran

Eryx miliaris (Pallas 1773)   *Anguis*
  Distribution: Iran and Afghanistan (north into U.S.S.R. and east to
            Mongolia)
  Reference: Steward (1971); Vogel (1964)

Eryx tataricus tataricus (Lichtenstein 1823)   *Boa*
  Distribution: Iran (and U.S.S.R.)

Eryx tataricus vittatus Chernov 1959
  Distribution: Eastern Iran, Afghanistan, (adjoining U.S.S.R. east to
            western China)
  Reference: Anderson and Leviton (1969)

Chapter 14

FAMILY COLUBRIDAE

Order: Serpentes
Suborder: Alethinophidia
Infraorder: Caenophidia
Superfamily: Colubroidea
Family: Colubridae

Subfamiy Aparallactinae

Genus: BRACHYOPHIS Mocquard 1888
Species typica: *revoili* Mocquard

Brachyophis revoili revoili Mocquard 1888
    Distribution: Yemen (and Somalia)
    Reference: Gasperetti (1977); Parker (1949); Witte and Laurent (1947)

Genus: MICRELAPS Boettger 1880
Species typica: *muelleri* Boettger
Reference: Witte and Laurent (1947)

Micrelaps muelleri Boettger 1880
    Distribution: Israel north to Syria

Subfamily Atractaspidinae

Genus: ATRACTASPIS Smith 1849
Species typica: *inornatus* Smith
Reference: Kochva, Shayer-Wollberg and Sobol (1967); Laurent (1950)

Atractaspis engaddensis Haas 1950
    Distribution: Israel, Jordan, Lebanon (and Egypt)
    Reference: Kochva (1959)

Atractaspis microlepidota andersoni Boulenger 1905
    Distribution: Southwestern Arabia

Subfamily Boiginae

Genus: BOIGA Fitzinger 1826
Species typica: *irregularis* Merrem

Boiga trigonota melanocephala Annandale 1904
    Distribution: east Iran, south and west Afghanistan
    Reference: Gans and Latifi (1973); Latifi, Hoge and Eliazan (1966);
        Minton (1962)

Genus: MALPOLON Fitzinger 1826
Species typica: *monspessulanus* Hermann

Malpolon miolensis (Reuss 1834)   *Coluber*
    Distribution: (N. Africa), Sinai and Israel east to Iran
    Reference: Domergue (1959); Schmidt and Inger (1957); Werner (1971)

Malpolon monspessulanus (Hermann)
    Reference: Arnold and Burton (1978); Darevsky (1956); Steward (1971)

Malpolon monspessulanus monspessulanus (Hermann 1804)   *Coluber*
    Distribution: northwest Italy, coastal France, Iberia (and northwest
        Africa)
    Reference: Dunson, Dunson and Keith (1978); Minton and Salanitro
        (1972)

Malpolon monspessulanus insignitus (Geoffroy St. Hilaire 1827)   *Coluber*
    Distribution: (N. Africa) east to Iran and north to Turkey; Yugoslavia,
        Albania, southern Bulgaria, Greece and Cyprus
    Reference: Gavish (1979); Werner (1971)

Genus: TELESCOPUS Wagler 1830
Species typica: 'Coluber on plate v.'

Telescopus dhara guentheri (Anderson 1895)   *Tarbophis*
    Distribution: Arabia
    Reference: Haas and Battersby (1959)

Telescopus dhara obtusus (Reuss 1834)   *Coluber*
    Distribution: (N. Africa), Sinai and Israel
    Reference: Marx (1968)

Telescopus fallax (Fleischmann)
    Reference: Kochva (1965); Steward (1971)

Telescopus fallax fallax (Fleischmann 1831)   *Tarbophis*
    Distribution: Balkan Peninsula and western Turkey

Telescopus fallax hoogstraali Schmidt and Marx 1956
    Distribution: Sinai
    Reference: Zinner (1977)

Telescopus fallax iberus (Eichwald 1831)   *Tarbophis*
    Distribution: eastern Turkey, northern Syria, Iraq, northwest Iran
        (and U.S.S.R.)

Telescopus fallax multisquamatus Wettstein 1952
    Distribution: Kufonisi Island, southeast of Crete

Telescopus fallax pallidus Stepanek 1944
    Distribution: Crete

Telescopus fallax syriacus Boettger 1896
    Distribution: southern Turkey south to Israel

Telescopus rhinopoma (Blanford 1874)   *Dipsas*
    Distribution: southern Iran, southern Afghanistan (and Pakistan)
    Reference: Bohme (1977)

Telescopus tessellatus tessellatus (Wall 1908)   *Tarbophis*
    Distribution: southwestern Iran and Iraq
    Reference: Anderson (1963)

Telescopus tessellatus martini (Schmidt 1939)   *Tarbophis*
    Distribution: Iraq

Subfamily Boodontinae

Genus: LAMPROPHIS Fitzinger 1843
Species typica: *aurora* Linnaeus

Lamprophis arabicus (Parker 1930)   *Boaedon*
    Distribution: southwest Arabia

Subfamily Colubrinae

Genus: CORONELLA Laurenti 1768
Species typica: *austriaca* Laurenti
Reference: Arnold and Burton (1978); Spellerberg (1976); Steward (1971)

Coronella austriaca Laurenti
    Reference: Andren and Nilsen (1976); Beebee (1978); Gabe and Saint
        Girons (1965); (1969); Jacobshagen (1937); Kozak and
        Simecek (1977); Pernkopf and Lehner (1937); Phelps (1978);
        Spellerberg and Phelps (1975)

Coronella austriaca austriaca Laurenti 1768
    Distribution: southern England and France east through central Europe
        and the Balkans to northern Turkey and northwest Iran
    Reference: Branch and Wade (1976); Spellerberg (1977); Spellerberg
        and Phelps (1977)

Coronella austriaca fitzingeri (Bonaparte 1840)   *Zacholus*
    Distribution: southern Italy and Sicily

Coronella girondica (Daudin 1803)   *Coluber*
    Distribution: Austria, Italy, Sicily, southern France, Iberia (and
        northwest Africa)
    Reference: Boger (1940); Pasteur and Bons (1960); Street (1973)

Genus: HAEMORRHOIS Boie 1826
Species typica: *hippocrepis* Linnaeus
Reference: Welch (1980)

Haemorrhois caudaelineata (Gunther 1858)   *Zamenis*
    Distribution: Iran

Haemorrhois elegantissimus (Gunther 1878)   *Zamenis*
  Distribution: Israel, Jordan, northwest and central Arabia
  Reference: Gasperetti (1977); Marx (1968)

Haemorrhois gemonensis (Laurenti 1768)   *Natrix*
  Distribution: Istria south to Greece; Corfu, Cythira, Crete and Euboia
  Reference: Steward (1971)

Haemorrhois hippocrepis hippocrepis (Linnaeus 1758)   *Coluber*
  Distribution: Iberia (and northwest Africa)
  Reference: Bruno and Hotz (1976); Steward (1971)

Haemorrhois jugularis (Linnaeus)
  Reference: Steward (1971)

Haemorrhois jugularis jugularis (Linnaeus 1758)   *Coluber*
  Distribution: Cyprus, southern Turkey, Syria, Lebanon, Israel, Jordan,
          Iraq and Iran
  Reference:
  Coluber jugularis asianus (Boettger): Haas and Werner (1969)

Haemorrhois jugularis caspius (Gmelin 1789)   *Coluber*
  Distribution: Hungary, Romania, Yugoslavia, Albania, Greece, Bulgaria,
          Turkey (and U.S.S.R.)
  Reference: Vogel (1964)

Haemorrhois jugularis schmidti (Nikolsky 1909)   *Coluber*
  Distribution: northwest Iran, extreme eastern Turkey (and U.S.S.R.)

Haemorrhois karelini (Brandt 1838)   *Coluber*
  Distribution: Afghanistan, eastern Iran (Pakistan and U.S.S.R.)
  Reference: Vogel (1964)

Haemorrhois najadum (Eichwald)
  Reference: Steward (1971)

Haemorrhois najadum najadum (Eichwald 1831)   *Tyria*
  Distribution: eastern Turkey, northern Iran (and U.S.S.R.)
  Reference: Ananjeva and Orlov (1977)

Haemorrhois najadum dahli (Schinz 1833)   *Coluber*
  Distribution: Yugoslavia and Bulgaria south through the Balkans and
          east through Turkey, Syria and Iraq to Israel; Cyprus

Haemorrhois najadum kalymnensis (Schneider 1980)   *Coluber*
  Distribution: Kalymnos Island, Aegean Sea

Haemorrhois ravergieri ravergieri (Menetries 1832)   *Coluber*
  Distribution: (Egypt), Israel north to Turkey and east to Afghanistan
      (and Pakistan)
  Reference: Anderson and Leviton (1969); Mamonov (1977); Vogel
      (1964)

Haemorrhois ravergieri cernovi (Mertens 1952)   *Coluber*
  Distribution: eastern Turkey (and U.S.S.R.)

Haemorrhois rhodorhachis rhodorhachis (Jan 1865)   *Zamenis*
  Distribution: (Libya and Somalia north through) Sinai to Syria and east
      to Afghanistan (and northwest India)
  Reference: Anderson (1963); Anderson and Leviton (1969); Haas and
      Werner (1969); Leviton and Anderson (1963); Minton
      (1962); Werner (1971)

Haemorrhois rogersi (Anderson 1893)   *Zamenis*
  Distribution: (Libya) north to Israel and Iraq
  Reference: Werner (1971)

Haemorrhois sinai (Schmidt and Marx 1956)   *Lytorhynchus*
  Distribution: Sinai

Haemorrhois thomasi (Parker 1931)   *Coluber*
  Distribution: southern Arabian Peninsula

Haemorrhois variabilis (Boulenger 1905)   *Zamenis*
  Distribution: southern Arabian Peninsula

Haemorrhois ventromaculatus (Gray 1834)   *Coluber*
  Distribution: Israel and eastern Arabia east to Pakistan
  Reference: Gasperetti (1977); Haas and Werner (1969); Minton (1962)

Haemorrhois viridiflavus (Lacepede)
  Reference: Steward (1971)

Haemorrhois viridiflavus viridiflavus (Lacepede 1789)   *Coluber*
  Distribution: Spain, France, Switzerland, Italy, Sardinia, Corsica and
      Elba

Haemorrhois viridiflavus carbonarius (Bonaparte 1833)   *Coluber*
    Distribution: southeast Austria, northeast Italy south through
                Yugoslavia; southern Italy, Sicily and Malta

Genus: LYTORHYNCHUS Peters 1862
Species typica: *diadema* Dumeril, Bibron and Dumeril
Reference: Leviton (1977); Leviton and Anderson (1970)

Lytorhynchus diadema (Dumeril, Bibron and Dumeril 1854)   *Heterodon*
    Distribution: Israel and Sinai (west to Morocco)
    Reference: Flower (1933); Schmidt and Inger (1957)

Lytorhynchus gaddi Nikolsky 1907
    Distribution: Arabian Peninsula, Iran, Iraq and Kuwait
    Reference:
    Lytorhynchus diadema arabicus Haas: Haas (1952)
    Lytorhynchus diadema mesopotamicus Haas: Haas (1952)

Lytorhynchus gasperetti Leviton 1977
    Distribution: Saudi Arabia

Lytorhynchus kennedyi Schmidt 1939
    Distribution: Syria

Lytorhynchus maynardi Alcock and Finn 1896
    Distribution: Afghanistan (and Pakistan)

Lytorhynchus ridgewayi Boulenger 1887
    Distribution: Iran, Afghanistan (and Pakistan)

Genus: MEIZODON Fischer 1856
Species typica: *coronatus* Schlegel

Meizodon semiornatus Peters 1854)   *Coronella*
    Distribution: Yemen (and eastern Africa)
    Reference: Gasperetti (1977)

Genus: PTYAS Fitzinger 1843
Species typica: *blumenbachii* Merrem = *mucosus* Linnaeus

Ptyas mucosus (Linnaeus 1758)   *Coluber*
  Distribution: Iran and Afghanistan (east through Asia)
  Reference: Anderson and Leviton (1969); Minton (1962); Schmidt and
        Inger (1957); Singh, Purdom and Jones (1976)

Genus: SPALEROSOPHIS Jan 1865
Species typica: *microlepis* Jan
Reference: Marx (1959)

Spalerosophis diadema cliffordi (Schlegel 1837)   *Coluber*
  Distribution: (North Africa) east through Syria, Arabia, Iraq to western
        Iran
  Reference: Dmi'el (1967); Dmi'el and Borut (1972); Dmi'el and Zilber
        (1971); Pasteur (1967)

Spalerosophis diadema schiraziana Jan 1865
  Distribution: Iran and Afghanistan
  Reference: Anderson and Leviton (1969); Minton (1962); Vogel (1964)

Spalerosophis microlepis Jan 1865
  Distribution: Iran

Subfamily Dasypeltinae

Genus: DASYPELTIS Wagler 1830
Species typica: *scaber* Linnaeus

Dasypeltis scabra (Linnaeus 1758)   *Coluber*
  Distribution: south Arabia (and Africa)
  Reference: Broadley (1966); Cogger (1966); Gabe and Saint Girons
        (1969); Gans (1952) (1959); Haas (1931a) (1931b) (1932);
        Pringle (1954); Rabb and Snediger (1960)

Subfamily Lycodontinae

Genus: LYCODON Fitzinger 1826
Species typica: *aulicus* Linnaeus

Lycodon striatus bicolor (Nikolsky 1903)   *Contia*
    Distribution: east and northeast Iran (and U.S.S.R.)

Subfamily Lycophidinae

Genus: LYCOPHIDION Fitzinger 1843
Species typica: *capense* Smith

Lycophidion capense jacksoni (Boulenger 1893)   *Lycophidium*
    Distribution: Yemen (and East Africa)
    Reference: Gasperetti (1977); Laurent (1968)

Genus: MACROPROTODON Guichenot 1850
Species typica: *cucullatus* Geoffroy

Macroprotodon cucullatus cucullatus (Geoffroy St. Hilaire 1827)   *Coluber*
    Distribution: Iberia and the Balearics; (North Africa east to) Sinai and
            Israel
    Reference: Gabe and Saint Girons (1965); Steward (1971)

Subfamily Natricinae

Genus: NATRIX Laurenti 1768
Species typica: *natrix* Linnaeus
Reference: Arnold and Burton (1978); Pasteur and Bons (1960); Steward
        (1971)

Natrix maura (Linnaeus 1758)   *Coluber*
    Distribution: southern France, western Switzerland, northwest Italy,
                Iberia, Balearics, Corsica, Sardinia (and northwest Africa)
    Reference: Bergmans (1976); Gabe (1972); Gabe and Saint Girons
            (1965) (1969) (1972)

Natrix natrix (Linnaeus)
>
> Reference: Armstrong (1951); Gabe (1971); Gabe and Saint Girons
> (1969); Jacobshagen (1937); Kostanecki (1926); Kuhnel and
> Krisch (1974); Landmann (1979); Mertens (1947); Saint
> Girons and Saint Girons (1956); Schnabel (1955) (1956);
> Schnabel and Herschel (1955); Sjongren (1945); Skoczylas
> (1970a) (1970b); Spellerberg (1976); Thorpe (1975a) (1975b)
> (1979) (1980a) (1980b)

Natrix natrix natrix (Linnaeus 1758)   *Coluber*
>
> Distribution: Scandinavia, Denmark and Germany east through Europe
> (to U.S.S.R.), south through the Balkans, Turkey, northern
> Syria, Iraq, and northern Iran; Cyprus
>
> Reference:
> *Natrix natrix gotlandica* Nilson and Andren: Nilson and Andren (1981a)
> (1981b); Thorpe (1981)
> *Natrix natrix persa* (Pallas): Mertens (1952a); Schulte (1974); Steward
> (1971)
> *Natrix natrix schweizeri* Muller: Muller (1933); Steward (1971)
> *Natrix natrix scutata* (Pallas): Mertens (1966b); Steward (1971)

Natrix natrix cetti Gene 1838
>
> Distribution: Sardinia

Natrix natrix corsa (Hecht 1930)   *Tropidonotus*
>
> Distribution: Corsica

Natrix natrix helvetica (Lacepede 1789)   *Coluber*
>
> Distribution: United Kingdom, France and Belgium south through Italy
> and Iberia; Sicily; and (north Africa)
> Reference: Branch and Wade (1976); Halfpenny and Bellairs (1976);
> Heusser and Schlumpf (1962); Petter-Rousseaux (1953)
> *Natrix natrix astreptophora* (Seoane): Pasteur and Bons (1960);
> Steward (1971)
> *Natrix natrix sicula* (Cuvier); Steward (1971)

Natrix tessellata tessellata (Laurenti 1768)   *Coronella*
>
> Distribution: Italy and Switzerland east through the Balkans to Turkey
> (and U.S.S.R.), south from Turkey to Sinai (and Egypt) and
> east to Afghanistan; Cyprus
> Reference: Anderson and Leviton (1969); Dummermuth (1977a);
> Gruschwitz (1978); Gygax (1971); Haas and Werner (1969);
> Jacobshagen (1937); Kostanecki (1926); Krapp and Bohme
> (1978); Steward (1958); Werner (1971)

Natrix tessellata heinrothi (Hecht 1930)   *Tropidonotus*
    Distribution: Serpilor Island near the Danube Delta

Genus: XENOCHROPHIS Gunther 1864
Species typica: *cerasogaster* Cantor

Xenochrophis piscator (Schneider 1799)   *Hydrus*
    Distribution: Afghanistan (south to Sri Lanka and east to China, Malaya
        and Indonesia)

Subfamily Philothamninae

Genus: PHILOTHAMNUS Smith 1847
Species typica: *semivariegatus* Smith

Philothamnus semivariegatus semivariegatus (Smith 1847)   *Dendrophis*
    Distribution: Yemen (and East Africa)
    Reference: Gasperetti (1977); Sweeny (1971)

Subfamily Psammophinae

Genus: PSAMMOPHIS Boie 1826
Species typica: *sibilans* Linnaeus

Psammophis leithii Gunther 1869
    Distribution: Afghanistan (Pakistan and India)
    Reference: Kral (1969)

Psammophis lineolatus (Brandt 1838)   *Coluber*
    Distribution: Iran, Afghanistan (and U.S.S.R.)

Psammophis schokari (Forskal 1775)   *Coluber*
    Distribution: North Africa east to Pakistan and India
    Reference: Anderson (1963); Haas and Werner (1969); Minton (1962);
        Minton and Salanitro (1972)

Subfamily 'uncertain'

Genus: EIRENIS Jan 1863
Species typica: *collaris* Menetries

Eirenis arabica Haas 1961
    Distribution: Arabia

Eirenis collaris (Menetries 1832)   *Coluber*
    Distribution: Jordan north to Turkey and Iran (and adjoining U.S.S.R.)
    Reference: Haas and Werner (1969); Steward (1971); Weber-Semenoff
        (1977)

Eirenis coronella coronella (Schlegel 1837)   *Calamaria*
    Distribution: Sinai, Israel, Iran, Iraq and Syria
    Reference: Haas and Werner (1969)

Eirenis coronella fraseri Schmidt 1939
    Distribution: Iraq

Eirenis coronelloides (Jan 1862)   *Homalosoma*
    Distribution: Sinai and Israel east to Iran

Eirenis decemlineata (Dumeril and Bibron 1854)   *Ablabes*
    Distribution: Israel east to Iran

Eirenis frenatus (Gunther 1858)   *Cyclophis*
    Distribution: Iran (east to India)

Eirenis iranica Schmidt 1939
    Distribution: Iran

Eirenis lineomaculata Schmidt 1939
    Distribution: Jordan Valley north to Turkey

Eirenis modestus modestus (Martin 1838)   *Coronella*
    Distribution: Cyprus; Turkey, south to Israel and east to Northwest
        Iran (and adjoining U.S.S.R.)
    Reference: Petack (1975); Steward (1971)

Eirenis modestus werneri Wettstein 1952
    Distribution: Alazonisi Rock near Furni Island off the coast of Turkey

Eirenis persicus (Anderson 1872)   *Cyclophis*
    Distribution: Turkey, Iran, Iraq, (adjoining U.S.S.R., Pakistan and
        northwest India)
    Reference: Anderson (1963); Haas and Werner (1969)

Eirenis rothii Jan 1863
Distribution: Israel east to Iran, north to Turkey

Genus: ELAPHE Fitzinger 1833
Species typica: *quatuorlineata* Lacepede
Reference: Steward (1971)

Elaphe dione (Pallas 1773)   *Coluber*
Distribution: northern Iran, Afghanistan (and adjoining U.S.S.R. east to
China and Korea)
Reference: Olexa (1975); Vogel (1964)

Elaphe hohenackeri hohenackeri (Strauch 1873)   *Coluber*
Distribution: north and eastern Turkey and northwest Iran (north into
U.S.S.R.)

Elaphe hohenackeri taurica Werner 1898
Distribution: central southern Turkey

Elaphe longissima longissima (Laurenti 1768)   *Natrix*
Distribution: western France and northeast Spain east through southern
and central Europe to western and northern Turkey and
northwest Iran (and adjoining U.S.S.R.)
Reference: Cattaneo (1975); Lotze (1975); Schmidt and Inger (1957)

Elaphe longissima persica Werner 1913
Distribution: northern Iran

Elaphe longissima rechingeri Werner 1932
Distribution: Amorgos Island, Cyclades

Elaphe longissima romana (Suckow 1798)   *Coluber*
Distribution: southern half of Italy and Sicily

Elaphe quatuorlineata quatuorlineata (Lacepede 1789)   *Coluber*
Distribution: Italy, Sicily, coastal Yugoslavia, Albania and western
Greece

Elaphe quatuorlineata muenteri Bedriaga 1881
Distribution: Myconos Island, Cyclades

Elaphe quatuorlineata praematura Werner 1935
Distribution: Ios Island, Cyclades

Elaphe quatuorlineata sauromates (Pallas 1814)   *Coluber*
 Distribution: northeast Greece, eastern Romania and Bulgaria (north
    into U.S.S.R.) east through Turkey to northwest Iran (and
    adjoining U.S.S.R.)
 Reference: Berthold (1976); Petzold (1976); Tiedemann and Haupl
    (1978)

Elaphe scalaris (Schinz 1822)   *Coluber*
 Distribution: Iberia and mediterranean France

Elaphe situla (Linnaeus 1758)   *Coluber*
 Distribution: southern Italy, Sicily, Malta, Balkan Peninsula, Crete and
    western and northern Turkey

Genus: OLIGODON Boie 1827
Species typica: *bitorquatus* Reinwardt

Oligodon arnensis (Shaw 1802)   *Coluber*
 Distribution: Afghanistan (east through India)
 Reference: Kral (1969)

Oligodon taeniolatus (Jerdon 1853)   *Coronella*
 Distribution: Afghanistan (?) (Pakistan south to Sri Lanka)

Genus: RHYNCHOCALAMUS Gunther 1864
Species typica: *melanocephalus* Jan

Rhynchocalamus arabicus Schmidt 1933
 Distribution: southwest Arabia

Rhynchocalamus melanocephalus (Jan)
 Reference: Darevsky (1970)

Rhynchocalamus melanocephalus melanocephalus (Jan 1862)   *Oligodon*
 Distribution: Israel and northwest Arabia north to Syria

Rhynchocalamus melanocephalus satunini (Nikolsky 1889)   *Contia*
 Distribution: southern Turkey, northern Iraq and Iran (and adjoining
    U.S.S.R.)

Chapter 15

FAMILY ELAPIDAE

Order: Serpentes
Suborder: Alethinophidia
Infraorder: Caenophidia
Superfamily: Elapsoidea
Family: Elapidae
Subfamily: Bungarinae
Tribe: Bungarini

Genus: BUNGARUS Daudin 1803
Species typica: *fasciata* Schneider

Bungarus caeruleus sindanus Boulenger 1897
    Distribution: Afghanistan (Pakistan and India)
    Reference: Kral (1969)

Tribe Najini

Genus: NAJA Laurenti 1768
Species typica: *naja* Linnaeus
Reference: Broadley (1968) (1974); Harding and Welch (1980)

Naja haje haje (Linnaeus 1758)   *Coluber*
    Distribution: (North and East Africa) north through Sinai to Jordan
    Reference: Behler and Brazaitis (1974); Hager (1905); Kopeyan and
                others (1973); Joubert and Taljaard (1978a) (1978b) (1978c);
                Miranda and others (1970); Mohamed and others (1975);
                Warrell, Barnes and Piburn (1976)

Naja haje arabica Scortecci 1932
    Distribution: west and south coasts of Arabia

Naja naja oxiana (Eichwald 1831)   *Tomyris*
    Distribution: Iran, Afghanistan (U.S.S.R., Pakistan and India)

Genus: WALTERINNESIA Lataste 1887
Species typica: *aegyptia* Lataste

Walterinnesia aegyptia Lataste 1887
    Distribution: Egypt, Israel, Syria, Lebanon, Jordan, Iraq, Iran, Kuwait
                and Saudi Arabia
    Reference: Anderson (1963); Gitter and others (1962); Haas and Werner
                (1969); Lee, Chen and Mebs (1976); Marx (1953); Schmidt
                and Inger (1957)

Chapter 16

FAMILY HYDROPHIIDAE

Order: Serpentes
Suborder: Alethinophidia
Infraorder: Caenophidia
Superfamily: Elapsoidea
Family: Hydrophiidae
Subfamily: Hydrophiinae
Tribe: Hydrophiini

Genus: HYDROPHIS Latreille 1802
Species typica: *fasciatus* Schneider

Hydrophis cyanocinctus Daudin 1803
    Distribution: Persian Gulf (east to the Philippines)
    Reference: Kasturirangen (1952); Liu and others (1973); Mishima and
        Okonogi (1962)

Hydrophis gracilis (Shaw 1802)   *Hydrus*
    Distribution: Persian Gulf (east to Australia)

Hydrophis lapemoides (Gray 1849)   *Aturia*
    Distribution: Persian Gulf (south to Sri Lanka)

Hydrophis ornatus ornatus (Gray 1842)   *Aturia*
    Distribution: Persian Gulf (east to Australia)
    Reference: Mittleman (1947); Rai (1969)

Hydrophis spiralis (Shaw 1802)   *Hydrus*
    Distribution: Persian Gulf (east to the Philippines)
    Reference: Bhatnagar (1957); Rai (1969)

Genus: LAPEMIS Gray 1835
Species typica: *curtus* Shaw

Lapemis curtus (Shaw 1802)   *Hydrus*
    Distribution: Persian Gulf (south along the west coast of India)

Genus: PELAMIS Daudin 1803
Species typica: *platurus* Linnaeus

Pelamis platurus (Linnaeus 1766)   *Anguis*
    Distribution: Persian Gulf (south to East Africa and east to Pacific
        America)
    Reference: Bolanos and others (1974); Graham (1974a) (1974b);
        Graham, Gee and Robison (1975); Graham, Rubinoff and
        Hecht (1971); Greene (1973); Hecht, Kropach and Hecht
        (1974); Hibbard and LaVergne (1972); Klawe (1964);
        Kropach (1971) (1972) (1975); Pickwell (1971); Shipman
        and Pickwell (1973); Tu, Lin and Bieber (1975); Zeiller
        (1969)

Genus: THALASSOPHIS Schmidt 1852
Species typica: *anomalus* Schmidt

Thalassophis viperinus Schmidt 1852
    Distribution: Persian Gulf (east to Indonesia)

Chapter 17

FAMILY VIPERIDAE

Order: Serpentes
Suborder: Alethinophidia
Infraorder: Caenophidia
Superfamily: Viperoidea
Family: Viperidae
Subfamily: Viperinae

Genus: BITIS Gray 1842
Species typica: *arietans* Merrem

Bitis arietans arietans (Merrem 1820)   *Vipera*
    Distribution: southwest Arabia (and Africa)
    Reference: Bolt and Ewer (1964); Broadley and Parker (1976); Howard
            (1975); Janecek (1976); Otis (1973); Parsons and Cameron
            (1977); Thomas (1972); Van der Walt (1972); Van der Walt
            and Joubert (1971) (1972); Warrell, Ormerod and Davidson
            (1975); Willemse, Hattingh and Coetzee (1979); Willemse
            and others (1979)

Genus: CERASTES Laurenti 1768
Species typica: *cerastes* Linnaeus
Reference: Hassan and El Hawary (1977); Labib, Halim and Farag (1979);
            Warburg (1964)

Cerastes cerastes cerastes (Linnaeus 1758)   *Coluber*
    Distribution: (North Africa) north through Sinai to Jordan, Lebanon,
        Iraq and Kuwait
    Reference: Mohamed, El-Serougi and Khaled (1969); Mohamed and
        Khaled (1966)

Cerastes cerastes gasperettii Leviton and Anderson 1967
    Distribution: Arabia

Cerastes vipera (Linnaeus 1758)   *Coluber*
    Distribution: Israel, Lebanon and Sinai (west across North Africa)
    Reference: Gabe and Saint Girons (1965); Jacobshagen (1937);
        Kostanecksi (1926)

Genus: ECHIS Merrem 1820
Species typica: *carinata* Schneider

Echis carinatus (Schneider)
    Reference: Anderson (1963); Bhattacharya and Gaitonde (1979);
        Cheymol and others (1973); Detrait, Izard and Boquet
        (1960); Edgar and others (1980); Izard and Boquet (1958);
        Kornalik (1963); Kornalik and Taborska (1972); Lefrou and
        Martignoles (1954); Morita, Iwanaga and Suzuki (1976);
        Parsons and Cameron (1977); Poguda (1972); Reid (1977);
        Schiek, Kornalik and Habermann (1972); Somani and Arora
        (1962); Taborska (1971); Theakston, Lloyd-Jones and Reid
        (1977); Warrell and Arnett (1976); Warrell and others (1974)
        (1977); Zimmermann, Habermann and Lasch (1971)

Echis carinatus pyramidum (Geoffroy St. Hilaire 1827)   *Scythale*
    Distribution: Afghanistan, Iran (and U.S.S.R.) west to North Africa
    Reference: Anderson and Leviton (1969)

Echis colorata Gunther 1878
    Distribution: Arabian Peninsula (and northeast Africa)
    Reference: Fainaru and others (1974); Gitter and others (1960);
        Mendelssohn (1965); Moav, Moroz and Vries (1963)

Echis multisquamatus Cherlin 1981
    Distribution: north and east Iran, southern Afghanistan (and U.S.S.R.)
    Reference: Cherlin and Tsellarius (1981)

Genus: ERISTOCOPHIS Alcock and Finn 1897
Species typica: *macmahoni* Alcock and Finn

Eristocophis macmahoni Alcock and Finn 1897
   Distribution: Iran, Afghanistan (and Pakistan)

Genus: PSEUDOCERASTES Boulenger 1896
Species typica: *persicus* Dumeril, Bibron and Dumeril

Pseudocerastes persicus persicus (Dumeril, Bibron and Dumeril 1854)
                                                              Cerastes
   Distribution: Iraq, Kuwait, Saudi Arabia, Iran, Afghanistan (and
          Pakistan)
   Reference: Anderson (1963)

Pseudocerastes persicus fieldi Schmidt 1930
   Distribution: Jordan, Israel and Lebanon
   Reference: Gitter and others (1962); Mendelssohn (1965)

Genus: VIPERA Laurenti 1768
Species typica: *aspis* Linnaeus
Reference: Saint Girons (1976) (1977); Saint Girons and Detrait (1978)

Vipera ammodytes (Linnaeus)
   Reference: Boulenger (1913); Bruno (1967); Kostanecki (1926); Jackson
          (1980); Richter and Vozenilek (1977); Saint Girons (1978);
          Thomas (1961)

Vipera ammodytes ammodytes (Linnaeus 1758)   *Coluber*
   Distribution: Austria, northern Italy, southern Czechoslovakia, western
          Hungary, Yugoslavia, southwestern Romania and north-
          western Bulgaria
   Reference: Nedjalkov and Nashkova (1961); Thomas (1969)

Vipera ammodytes meridionalis Boulenger 1903
   Distribution: Albania, Greece, Cyclades, Turkey, Syria and Lebanon

Vipera ammodytes montandoni Boulenger 1904
   Distribution: Romania and Bulgaria
   Reference: Thomas (1971)

Vipera aspis (Linnaeus)
Reference: Acher, Chauvet and Chauvet (1968); Agid and others
(1961a) (1961b); Baumann (1929); Beguin (1902); Boffa and
Boffa (1971) (1974); Bruno (1976); Buresch and Beskov
(1965); Castanet (1974); Castanet and Nolleau (1974); Cesari
and Boquet (1939); Dastugue and Joy (1943); Detrait and
Duguy (1966); Duguy (1962); Duguy and Saint Girons
(1969); Gabe (1969) (1971) (1972); Gabe and Saint Girons
(1962) (1965) (1969) (1972); Gabriel-Robez and Clavert
(1980); Gonzalez (1976); Hubert (1966) (1968a); Izard,
Detrait and Boquet (1961); Naulleau (1965) (1966) (1967)
(1968) (1970) (1971) (1973a) (1973b) (1973c) (1976) (1979);
Peyer (1912); Saint Girons (1957) (1959) (1978); Saint
Girons and Duguy (1962); Schweizer (1921)

Vipera aspis aspis (Linnaeus 1758)   *Coluber*
Distribution: France, Switzerland, southern Austria, Germany, Italy and
northwestern Yugoslavia
Reference: Dummermuth (1977a)
Vipera aspis francisciredi Laurenti: Blattler (1976); Harding and Welch
(1980)

Vipera aspis atra Meisner 1820
Distribution: Switzerland

Vipera aspis hugyi Schinz 1833
Distribution: southern Italy and Sicily

Vipera aspis montecristi Mertens 1956
Distribution: Monte Cristo Island

Vipera aspis zinnikeri Kramer 1958
Distribution: southwestern France

Vipera berus (Linnaeus)
Reference: Bellairs, Griffiths and Bellairs (1955); Belova (1975) (1976);
Dullemeijer (1956); Jacobshagen (1937); Kostanecki (1926);
Jenkins and Simkiss (1968); Marshall and Woolf (1957);
Nilson (1976) (1980); Saint Girons (1978); Saint Girons and
Kramer (1963); Sjongren (1945); Spellerberg (1976); Thomas
(1955); Vainio (1931); Vladescu (1965a); Volsoe (1944);
Wolter (1924)

Vipera berus berus (Linnaeus 1758)   *Coluber*
    Distribution: Europe from the Arctic Circle south to the Pyrenees and
        from the United Kingdom east through Asia to the Amur
        River
    Reference: Bernstrom (1943); Branch and Wade (1976); Marian (1963);
        Phelps (1978); Prestt (1971); Reid (1976); Simms (1972);
        Viitanen (1967)

Vipera berus bosniensis Boettger 1881
    Distribution: Yugoslavia and Bulgaria

Vipera bornmuelleri Werner 1898
    Distribution: Lebanon
    Reference: Nilson and Sundberg (1981)

Vipera kaznakovi Nikolsky 1909
    Distribution: northeast Turkey (and U.S.S.R.)
    Reference: Kramer (1961); Mertens (1952a) (1952b); Saint Girons
        (1978); Vogel (1964)

Vipera latastei Bosca
    Reference: Bernis (1968); Saint Girons (1978)

Vipera latastei latastei Bosca 1878
    Distribution: Portugal, central and eastern Spain

Vipera latastei gaditana Saint Girons 1977
    Distribution: southern Spain (and northwest Africa)

Vipera latifii Mertens, Darevsky and Klemmer 1967
    Distribution: Iran
    Reference: Andren and Nilson (1979); Nilson and Sundberg (1981)

Vipera lebetina (Linnaeus)
    Reference: Cesari and Boquet (1937); Korneva (1973)

Vipera lebetina lebetina (Linnaeus 1758)   *Coluber*
    Distribution: Cyprus
    Reference: Bodenheimer (1957)

Vipera lebetina euphratica Martin 1838
    Distribution: Iraq
    Reference: Haas and Werner (1969)

Vipera lebetina obtusa Dwigubsky 1832
   Distribution: Turkey, Iraq, Afghanistan, Syria, Israel, Lebanon, Iran
      (Pakistan and U.S.S.R.)

Vipera lebetina schweizeri Werner 1935
   Distribution: Cyclades
   Reference: Mertens (1955); Stemmler (1967); Zingg (1968); Zwinenberg
      (1979)

Vipera lebetina turanica Chernov 1940
   Distribution: northeast Iran, Afghanistan (Pakistan, Kashmir and
      U.S.S.R.)
   Reference: Bogdanov and Zinyakova (1965)

Vipera palaestinae Werner 1938
   Distribution: Syria, Jordan, Israel and Lebanon
   Reference: Allon and Kochva (1974); Frenkel and Kochva (1970);
      Gitter and others (1957); Grotto and others (1967); Izard
      and Boquet (1958); Kochva (1958) (1960) (1961) (1962)
      (1963); Kochva and Gans (1965); Kochwa and others
      (1959) (1960); Mendelssohn (1963); Moroz, Goldblum and
      Vries (1963) (1965); Moroz, Hahn and Vries (1971); Moroz,
      Vries and Sela (1966); Nilson and Sundberg (1981); Ovadia,
      Kochva and Moav (1977); Rotenberg, Bamberger and
      Kochva (1971); Shaham and Kochva (1969)

Vipera raddei Boettger 1890
   Distribution: eastern Turkey, northwest Iran (and U.S.S.R.)
   Reference: Mertens (1952a); Nilson and Sundberg (1981); Vogel (1964)

Vipera seoanei Lataste 1879
   Distribution: north Portugal and northwest Spain
   Reference: Bea (1978a); Saint Girons (1978); Saint Girons and Duguy
      (1976)

Vipera ursinii (Bonaparte)
   Reference: Jacobshagen (1937); Kostanecki (1926); Kramer (1961);
      Saint Girons (1978)

Vipera ursinii ursinii (Bonaparte 1835)   *Pelias*
   Distribution: southeast France, central Italy, western Yugoslavia and
      north Albania; western Turkey

Vipera ursinii ebneri Knoepffler and Sochurek 1955
 Distribution: northeast Turkey, northwest Iran (and U.S.S.R.)

Vipera ursinii rakosiensis Mehely 1894
 Distribution: eastern Austria, Hungary, northern Yugoslavia, southern
   Romania and northern Bulgaria

Vipera ursinii renardi (Cristoph 1861)   *Pelias*
 Distribution: northeast Romania (and U.S.S.R.)

Vipera xanthina (Gray 1849)   *Daboia*
 Distribution: western Turkey
 Reference: Mertens (1952a); Nilson and Sundberg (1981); Vogel (1964)

Chapter 18

FAMILY CROTALIDAE

Order: Serpentes
Suborder: Alethinophidia
Infraorder: Caenophidia
Superfamily: Viperoidea
Family: Crotalidae
Subfamily: Crotalinae

Genus: AGKISTRODON Beauvois 1799
Species typica: *mokasen* Beauvois

Agkistrodon halys caucasicus Nikolsky 1916
    Distribution: northern Iran (and U.S.S.R.)

Agkistrodon himalayanus (Gunther 1864)   *Halys*
    Distribution: eastern Afghanistan (and Pakistan east through the
        Himalayas)

FAMILIAL REFERENCES

Trogonophidae: Gans (1967)
Amphisbaenidae: Gans (1967)
Gekkonidae: Underwood (1954); Wermuth (1965)
Agamidae: Wermuth (1967)
Chamaeleonidae: Mertens (1966)
Scincidae: Greer (1970) (1979)
Lacertidae: Boulenger (1920) (1921)
Anguidae: Wermuth (1969)
Varanidae: Mertens (1963)
Leptotyphlopidae: Hahn (1978a) (1978b); List (1966); McDowell (1967)
Typhlopidae: Hahn (1978a); List (1966); McDowell (1967) (1974)
Boidae: McDowell (1975) (1979); Stimson (1969)
Colubridae: Bourgeois (1968); Underwood (1967)
Atractaspidinae: Harding and Welch (1980); Laurent (1950)
Elapidae: Harding and Welch (1980)
Hydrophiidae: Burger and Natsuno (1974); Harding and Welch (1980);
            Smith (1926); Voris (1977)
Viperidae: Harding and Welch (1980); Marx and Rabb (1965)
Crotalidae: Harding and Welch (1980)

Appendix 2

REGIONAL REFERENCES

ABU DHABI: Leviton and Anderson (1967), see also Arabia; AFGHANISTAN: Anderson and Leviton (1969), Bruck (1968), Casimir (1970), Clark and others (1969), Kral (1969), Levitón (1959), Leviton and Anderson (1961) (1963) (1970b); ALBANIA: Kopstein and Wettstein (1920); ARABIAN PENINSULA: Corkill and Cochrane (1966), Gasperetti (1977), Haas (1957) (1961), Haas and Battersby (1959), Haas and Werner (1969), Schmidt (1930) (1941); AUSTRIA: Eiselt (1961); BELGIUM: Witte (1948); BULGARIA: Beskov and Beron (1964); CORFU: Mertens (1961); CYPRUS: Clarke (1973); CZECHOSLOVAKIA: Lac (1968); EGYPT (Sinai Peninsula): Barbour (1914), Marx (1968), Schmidt and Marx (1956); EUROPE: Arnold and Burton (1978), Mertens and Wermuth (1960), Steward (1971); FINLAND: Langerwert (1975); FRANCE: Angel (1946), Fretey (1975), Rollinat (1934); GREECE: Werner (1938), Wettstein (1953) (1957); HUNGARY: Fejervary-Langh (1943); IRAN: Anderson (1963) (1974), Casimir (1970), Clark, Clark and Anderson (1966), Forcart (1950), Guibe (1957) (1966), Haas and Werner (1969), Hellmich (1959), Latifi, Hoge and Eliazan (1966), Mertens (1956), Schleich (1977), Schmidt (1955), Tuck (1971), Werner (1903) (1938), Wettstein (1951); ITALY: Bruno (1970), Capocaccia (1968), Tortonese and Lanza (1968); IRAQ: Haas (1952c), Haas and Werner (1969), Khalaf (1959), Reed and Marx (1959); ISRAEL: Barash and Hoofien (1956), Hoofien (1972); LEBANON: Muller and Wettstein (1933), Zinner (1967), MONTECHRISTO: Bruno (1975); NETHERLANDS: Bund (1964); POLAND: Berger, Jaskowska and Mlynarski (1969); PORTUGAL: Crespo (1972); RHODES: Mertens (1959), Tortonese (1948), Wettstein (1964); SPAIN: Mattison and Smith (1978); SWEDEN: Gislen and Kauri (1959); SWITZERLAND: Grossenbacher and Brand (1973); SYRIA: Angel (1936), Barbour (1914), Haas and Werner (1969), Werner (1939), Wettstein (1928); TURKEY: Andren and Nilson (1976b), Bird (1936), Bodenheimer (1944), Clark (1972), Mertens (1952a) (1952b), Venzmer (1919), Werner (1903); UNITED KINGDOM: Smith (1973); YEMEN: Arillo (1967), Schmidt (1953), Scortecci (1932), see also Arabia; YUGOSLOVIA: Pozzi (1966)

BIBLIOGRAPHY

Acher, R., Chauvet, J. and Chauvet, M. T. (1968). Les hormones neurohypophysaires des reptiles: Isolement de la mesotocine et de la vasotocine de la vipere *Vipera aspis*. *Biochim. biophys. Acta Amsterdam 154,* 255-257.

Agid, R., Duguy, R., Martoja, M. and Saint Girons, H. (1961). Influence de la temperature et des facteurs endocrines dans la glycoregulation chez *Vipera aspis*. Role de l' adrenaline. *C. R. hebd. Seanc. Acad. Sci. Paris 252,* 2007-2009.

Agid. R., Duguy, R. and Saint Girons, H. (1961). Variations de la glycemie du glycogene hepatique et de l'aspect histologique du pancreas, chez *Vipera aspis,* au cours du cycle annuel. *J. Physiol. Paris 53,* 807-824.

Alexander, A. A. (1966). Taxonomy and variation of *Blanus strauchi,* with comments on the nature of its meristic variation. *Copeia* 205-224.

Ali, S. M. (1950). Studies on the comparative anatomy of the tail in Sauria and Rhynchocephalia. IV. *Anguis fragilis. Proceedings of the Indian Academy of Science 32,* 87-95.

Allen, A. (1977) Changes in population density of the eyed lizard, *Lacerta lepida,* at three localities in Portugal between 1969 and 1975. *British Journal of Herpetology 5,* 661-662.

Allon, N. and Kochva, E. (1974). The quantities of venom injected into prey of different size by *Vipera palaestinae* in a single bite. *Journal of experimental zoology 188,* 71-76.

Al-Nassar, N. A. (1976). Anatomical studies. Osteology and gut histology of the amphisbaenian *Diplometopon zarudnyi* inhabiting Kuwait. *Thesis M. Sc., University of Kuwait.*

Ananjeva, N. B. (1977). Morphometrical analysis of limb proportions of five sympatric species of desert lizards (Sauria, *Eremias*) in the southern Balkhash Lake Region. *Proceedings of the Zoological Institute of the Academy of Sciences, U.S.S.R. 74,* 3-13 (in Russian).

Ananjeva, N. B. (1981). Structural characteristics of skull, dentition and hyoid of lizards of the genus *Agama* from the fauna of the U.S.S.R., *Proceedings of the Zoological Institute of the Academy of Sciences, U.S.S.R. 101,* 3-20 (in Russian)

Ananjeva, N. B. and Orlov, N. L. (1977). On the records of the snake *Coluber najadum* (Eichw.) in southwestern Turkmenia. *Proceedings of the Zoological Institute of the Academy of Sciences, U.S.S.R. 74,* 14-16 (in Russian).

Ananjeva, N. B. and Orlova, V. F. (1979). Distribution and geographic variability of *Agama caucasia* (Eichwald, 1831). *Proceedings of the Zoological Institute of the Academy of Sciences, U.S.S.R. 89,* 4-17 (in Russian).

Ananjeva, N. B., Peters, G. and Rzepakovsky, V. T. (1981). New species of the mountain agamas from Tadjikistan, *Agama chernovi* sp. nov. *Proceedings of the Zoological Institute of the Academy of Sciences, U.S.S.R. 101,* 23-27 (in Russian).

Anderson, J. H. (1971). Utbredelse av firfisle, *Lacerta vivipara,*: Troms fylke. *Fauna 24,* 38-40.

Anderson, S. C. (1961). A note on the synonymy of *Microgecko* Nikolsky with *Tropiocolotes* Peters. *Wasmann Journal of Biology 19*, 287-289.

Anderson, S. C. (1963). Amphibians and reptiles from Iran. *Proceedings of the California Academy of Science (ser. 4) 31*, 417-498.

Anderson, S. C. (1966a). A substitute name for *Agama persica* Blanford. *Herpetologica 22*, 230.

Anderson, S. C. (1966b). The lectotype of *Agama isolepis* Boulenger. *Herpetologica 22*, 230-231.

Anderson, S. C. (1973). A new species of *Bunopus* (Reptilia: Gekkonidae) from Iran and a key to the lizards of the genus *Bunopus*. *Herpetologica 29*, 355-358.

Anderson, S. C. (1974). Preliminary key to the turtles, lizards and amphisbaenians of Iran. *Fieldiana Zoology 65*, 27-44.

Anderson, S. C. and Leviton, A. E. (1966a). A new species of *Eublepharis* from southwestern Iran. *Occasional Papers of the California Academy of Science 53*, 1-5.

Anderson, S. C. and Leviton, A. E. (1966b). A review of the genus *Ophiomorus* (Sauria: Scincidae) with descriptions of three new forms. *Proceedings of the California Academy of Sciences (ser. 4) 33*, 499-534.

Anderson, S. C. and Leviton, A. E. (1967a). A new species of *Eremias* (Sauria: Lacertidae) from Afghanistan. *Occasional Papers of the California Academy of Sciences 64*, 1-4.

Anderson, S. C. and Leviton, A. E. (1967b). A new species of *Phrynocephalus* (Sauria: Agamidae) from Afghanistan, with remarks on *Phrynocephalus ornatus* Boulenger. *Proceedings of the California Academy of Sciences (ser. 4) 35*, 227-234.

Anderson, S. C. and Leviton, A. E. (1969). Amphibians and reptiles collected by the Street Expedition to Afghanistan, 1965. *Proceedings of the California Academy of Sciences (ser. 4) 37*, 25-56.

Andren, C. and Nilsen, G. (1976a). Hasselsnoken *(Coronella austriaca)* en Utrotning-shobad Ormat. *Foch och Flora 71*, 61-76.

Andren, C. and Nilsen, G. (1976b). Observations on the herpetofauna of Turkey in 1968-1973. *British Journal of Herpetology 5*, 575-584.

Andren, C. and Nilsen, G. (1979). *Vipera latifii* (Reptilia, Serpentes, Viperidae) an endangered viper from Lar Valley, Iran, and remarks on the sympatric herpeto-fauna. *Journal of Herpetology 13*, 335-341.

Angel, F. (1936). Reptiles et batraciens de Syrie et de Mesopotamie recoltes par Paul Pallary. *Bulletin of the Institute of Egypt 18*, 107-116.

Angel, F. (1946). *Faune de France. Reptiles et Amphibiens*. Lechavalier, Paris.

Angel, F. and Lhote, H. (1938). Reptiles et Amphibiens du Sahara central et du Soudan. *Bull. Com. Et. hist. sci. Afr. occ. franc. 21*, 345-384.

Appleyard, D. C. (1978). Notes on the care and breeding of green lizards *(Lacerta viridis)*. *British Herpetological Society Newsletter 19*, 12-14.

Appleyard, D. C. (1979a). Care of wall lizards *(Podarcis muralis)*. *Herptile 4*, 33-34.

Appleyard, D. C. (1979b). Notes on the incubation of four clutches of green lizard *(Lacerta viridis)* eggs. *Herptile 4*, 35-38.

Arillo, A. (1967). Missione Scortecci 1965 nello Yemen: Reptilia, Testudines. *Boll. Musei Ist. Biol. Univ. Genova 35*, 185-192.

Armstrong, J. A. (1950). An experimental study of the visual pathways in a reptile *(Lacerta vivipara)*. *Journal of Anatomy 84*, 146-167.

Armstrong, J. A. (1951). An experimental study of the visual pathways in a snake *(Natrix natrix)*. *Journal of Anatomy 85*, 275-288.

Arnold, E. N. (1972). Lizards with northern affinities from the mountains of Oman. *Zool. Mede. Leiden 47*, 111-128.

Arnold, E. N. (1973). Relationships of the Palaearctic lizards assigned to the genera *Lacerta, Algyroides* and *Psammodromus* (Reptilia: Lacertidae). *Bulletin of the British Museum (Natural History), Zoology 25,* 289-366.

Arnold, E. N. and Burton, J. A. (1978). *A field guide to the reptiles and amphibians of Britain and Europe.* Collins, London 272 p.

Arnold, E. N. and Leviton, A. E. (1977). A revision of the lizard genus *Scincus* (Reptilia: Scincidae). *Bulletin of the British Museum (Natural History), Zoology 31,* 187-248.

Arronet, V. N. (1973). Morphological changes of nuclear structures in the Oogenesis of reptiles (Lacertidae, Agamidae). *Journal of Herpetology 7,* 163-193.

Asana, J. (1931). The natural history of *Calotes versicolor,* the common blood sucker. *Journal of the Bombay natural history society 34,* 1041-1047.

Auffenberg, W. (1979). Intersexual differences in behavior of captive *Varanus bengalensis* (Reptilia, Lacertilia, Varanidae). *Journal of Herpetology 13,* 313-315.

Avery, R. A. (1962). Notes on the ecology of *Lacerta vivipara. British Journal of Herpetology 3,* 36-39.

Avery, R. A. (1966). Food and feeding habits of the common lizard *(Lacerta vivipara)* in the west of England. *Journal of Zoology, London 149,* 115-121.

Avery, R. A. (1970). Utilisation of caudal fat by hibernating common lizards *Lacerta vivipara. Comparative Biochemistry and Physiology 37,* 119-121.

Avery, R. A. (1971). Estimates of food consumption by the lizard *Lacerta vivipara* Jacquin. *Journal of animal ecology 40,* 351-365.

Avery, R. A. (1973). Morphometric and functional studies on the stomach of the lizard *Lacerta vivipara. Journal of Zoology, London 169,* 157-167.

Avery, R. A. (1975a). Age structure and longevity of common lizard *(Lacerta vivipara)* populations. *Journal of Zoology, London 176,* 555-558.

Avery, R. A. (1975b). Clutch size and reproductive effort in the lizard *Lacerta vivipara* Jacquin. *Oecologia (Berlin) 19,* 165-170.

Avery, R. A. (1976). Thermoregulation, metabolism and social behaviour in Lacertidae In, Bellairs, A. d'A and Cox, C. B. (Editors), *Morphology and Biology of Reptiles.* Academic Press. 245-259.

Avery, R. A. (1978). Activity patterns, thermoregulation and food consumption in two sympatric lizard species *(Podarcis muralis* and *'P. sicula)* from central Italy. *Journal of Animal Ecology 47,* 143-158.

Avery, R. A. and McArdle, B. H. (1973). The morning emergence of the common lizard *Lacerta vivipara* Jacquin. *British Journal of Herpetology 5,* 363-368.

Avery, R. A., Shewry, P. R. and Stobart, A. K. (1974). A comparison of lipids from the fat body and tail of the common lizard, *Lacerta vivipara. British Journal of Herpetology 5,* 410-412.

Baby, T. G. and Reddy, S. R. R. (1977). Nitrogenous constituents in the urinary deposits of the lizard *Calotes versicolor. British Journal of Herpetology 5,* 649-653.

Bachmann, M. (1979). Jagd auf *Lacerta muralis. Aquar. Terr., Berlin (Ost) 26,* 194-195.

Badir, N. (1958). Seasonal variation of the male urogenital organs of *Scincus scincus* L. and *Chalcides ocellatus* Forsk. *Z. wiss. Zool. 160,* 290-351.

Badir, N. (1968a). Structure and function of corpus luteum during gestation in the viviparous lizard *Chalcides ocellatus. Anat. Anz. 122,* 1-10.

Badir, N. (1968b). The effect of population density on the embryonic mortality in the viviparous lizard *Chalcides ocellatus* (Forsk.) *Anat. Anz. 122,* 11-14.

Badir, N. and Hussein, M. F. (1965). Effect of temperature, food and illumination on the reproduction of *Chalcides ocellatus* (Forsk.) and *Scincus scincus* (Linn.) *Bulletin of the Faculty of Science, Cairo University 39,* 179-185.

Baecker, R. (1940). Uber die als Stratum fibrosum (compactum) bezeichnete Grenzschicht im Verdauungs-Kanal der Wirbeltiere. *Z. Forsch. mikrosk. Anat.* 47, 49-99.

Bailey, S. E. R. (1969). The responses of sensory receptors in the skin of the green lizard, *Lacerta viridis,* to mechanical and thermal stimualtion. *Comparative Biochemistry and Physiology 29,* 161-172.

Baker, J. R. (1942). The free border of the intestinal epithelial cell of vertebrates. *Q. J. microsc. Sci. 84,* 73-103.

Bannister, L. H. (1968). Fine structure of the sensory endings in the vomero-nasal organ of the slow worm *Anguis fragilis. Nature, London 217,* 275-276.

Baranoff, A. S. and Valetzky, A. V. (1975). Geographical variability of the body weight in the sand lizard *(Lacerta agilis)* in the U.S.S.R. *Zool. Zhur. 54,* 789-791 (in Russian).

Barash, A. and Hoofien, J. H. (1956). *Reptiles of Israel.* Tel-Aviv, Hakibuts Hameuchat, 180 pp.

Barbour, T. (1914). Notes on some reptiles from Sinai and Syria. *Proceedings of the New England Zoological Club 5,* 73-92.

Baumann, F. (1929). Experimente uber den Geruchssinn und den Beuteerwerb der Viper *(Vipera aspis* L.). *Z. Physiol. 10,* 36-119.

Baur, B. (1979a). Weisse Scheibenfinger — *Hemidactylus turcicus. Das Aquarium (Minden) 13,* 128-131.

Baur, B. (1979b). Unauffallig hubsch — die Waldeidechse, *Lacerta vivipara* Jacquin, 1787. *Das Aquarium (Minden) 13,* 271-273.

Bea, A. (1978a). Contribucion a la sistematica de *Vipera seoanei* Lataste, 1879 (Reptilia, Viperidae) 1. Ultrastructura de la cuticula de las escamas. *Butll. Inst. Cat. Hist. Nat. 42 (Sec. Zool., 2)* 107-118.

Bea, A. (1978b). Nota sobre *Lacerta vivipara* Jacquin, 1787, en la Peninsula Iberica. *Butll. Inst. Cat. Hist. Nat. 42 (Sec. Zool., 2)* 123-126.

Beebee, T. J. C. (1978). An attempt to explain the distributions of the rare herptiles *Bufo calamita, Lacerta agilis* and *Coronella austriaca* in Britain. *British Journal of Herpetology 5,* 763-770.

Beguin, R. (1902). Contribution a l'etude histologique du tube digestif des reptiles. *Revue suisse zoologique 10,* 250-397.

Behler, J. L. and Brazaitis, P. (1974). Breeding the Egyptian cobra, *Naja haje,* at the New York Zoological Park. *International Zoo Yearbook 14,* 83-84.

Bellairs, R., Griffiths, I. and Bellairs, A. d'A. (1955). Placentation in the adder *Vipera berus. Nature, London 176,* 657-658.

Belova, Z. V. (1975). Sex and composition of the *Vipera berus* population. *Zool. Zhur. 54,* 143-145 (in Russian).

Belova, Z. V. (1976). Spatial structure of a population of the common adder *(Vipera berus). Ekologiya 1,* 71-75 (in Russian).

Berger, L., Jaskowska, J. and Mlynarski, M. (1969). Plazy i gadi. *Katalogue Fauny Polski 39,* 1-73.

Bergmans, H. (1976). Die Vipernatter, *Natrix maura. Aquarien Terrarien Z. 29,* 318-320.

Bernis, F. (1968). La culebra de las islas Columbretes: *Vipera lastastei. Bol. R. Soc. espan. Hist. nat. 66,* 115-133.

Bernstrom, J. (1943). Till kannedom om huggormen *Vipera berus berus* (Linne). *Meddn. Goteborgs Mus. Zool. Avd. 103,* 1-34.

Berthold, M. (1976). Beobachtungen bei *Elaphe quatuorlineata sauromates. Aquar. Terr. 23,* 314.

Beskov, V. and Beron, P. (1964). *Catalogue et Bibliographie des Amphibiens et des Reptiles en Bulgarie.* Bulgarian Academy of Sciences, Sofia.

Beutler, A. and Gruber, U. (1979). Geschlechtsdimorphismus populationsdynamik und okologie von *Cyrtodactylus kotschyi* (Steindachner, 1870) (Reptilia: Sauria: Gekkonidae). *Salamandra 15*, 84-95.

Bhatnagar, H. M. (1957). The fronto-parietal region in the skull of *Hydrophis spiralis* Shaw. *Science and Culture 23*, 196.

Bhattacharya, D. R. (1921). Notes on the venous system of *Varanus bengalensis*. *Journal of the Proceedings of the Asiatic Society of Bengal 17*, 257-261.

Bhattacharya, S. and Gaitonde, B. B. (1979). Partial purification of cholinesterase from the venom of the saw scaled viper *(Echis carinatus)*. *Toxicon 17*, 429-431.

Bhattacharya, S. and Ghose, K. C. (1970). Occurrence of biliary amylase in vertebrates: influence of NaCl and pH. *Comparative Biochemistry and Physiology 37*, 581-587.

Bird, C. G. (1936). The distribution of reptiles and amphibians in Asiatic Turkey with notes on a collection from the Vilayets of Adana, Gaziantep and Malatya. *Ann. Mag. Nat. Hist. (ser. 10) 18*, 257-281.

Bischoff, W. (1974a). Echsen des Kaukasus, Teil 4. Die Artwiner Eidechse, *Lacerta derjugini* Nikolski 1898. *Aquar. Terr. B21*, 63-66.

Bischoff, W. (1974b). Echsen des Kaukasus, Teil 5. Die Kaukasus-Riesensmaragdeidechse, *Lacerta trilineata media* Lantz and Cyren 1920. *Aquar. Terr. B21*, 114-117.

Bischoff, W. (1974c). Echsen des Kaukasus, Teil 6. Die Kielschwanz-Felseidechse *Lacerta rudis* Bedriaga 1886. *Aquar. Terr. B21*, 274-278.

Bischoff, W. (1974d). Echsen des Kaukasus, Teil 7. Die Europaische Schlangenaugen-Eidechse *Ophisops elegans* Menetries 1832. *Aquar. Terr. B21*, 340-343.

Bischoff, W. (1975a). Echsen des Kaukasus, Teil 8. *Aquar. Terr. B22*, 51-53.

Bischoff, W. (1975b). Erganzende Mitteilungen zur Verbreitung von *Lacerta trilineata media*. *Aquar. Terr. B22*, 103.

Bischoff, W. (1975c). Echsen des Kaukasus, Teil 9. Die Bastardeidechse. *Aquar. Terr. B22*, 230-232.

Bischoff, W. (1976a). Echsen des Kaukasus, Teil 10. Die Streifeneidechse *Lacerta strigata* Eichwald 1831. *Aquar. Terr. B23*, 84-88.

Bischoff, W. (1976b). Echsen des Kaukasus, Teil 11. Die Wieseneidechse *Lacerta praticola* Eversmann 1834. *Aquar. Terr. B23*, 415-417.

Blanc, C. P. (1979). Notes sur les Reptiles de Tunisie: VI.—Observations sur la morphologie et les biotopes des *Mesalina* (Reptilia: Lacertidae). *C. R. Soc. Biogeogr. 491*, 53-61.

Blanchard, R. (1890). Sur une remarquable dermatose causee chez le lezard vert par un champignon du genre *Selenosporium*. *Memoirs Societe Zoologique de France 3*, 241-255.

Blasco, M. (1979). *Chamaeleo chamaeleon* in the province of Malaga, Spain. *British Journal of Herpetology 5*, 839-841.

Blattler, E. (1976). Die Trachtigkeit einer *Vipera aspis francisciredi* im Terrarium. *Aquaria 23*, 133-135.

Boag, D. A. (1973). Spatial relationships among members of a population of wall lizards. *Oecologia 12*, 1-13.

Bodenheimer, F. S. (1944). Introduction into the knowledge of the amphibia and reptilia of Turkey. *Revue de la Faculte des Sciences, University of Istanbul (ser. B) 9*, 1-193.

Bodenheimer, F. S. (1957). *Vipera lebetina lebetina* Linnaeus 1758 in Palestine. *Studies in Biology* (Jerusalem) *1*, 114-118.

Boettger, O. (1888). Verzeichnis der von Herrn E. v. OERTZEN aus Griechenland und Kleinasien mitgebrachten Batrachier und Reptilien. *Sitzb. Akad. Wiss. Berlin 5*, 139-186.

Boffa, M. C. and Boffa, G. A. (1971). Identification et separation de differents facteurs du venin de *Vipera aspis* actifs en hemostase. *C. R. Seanc. Soc. Biol.* *156,* 2287-2293.

Boffa, M. C. and Boffa, G. A. (1974). Correlations between the enzymatic activities and the factors active on blood coagulation and platelet aggregation from the venom of *Vipera aspis*. *Biochim biophys. Acta 354,* 275-290.

Bogdanov, O. P. and Vashetko, E. V. (1972). On ecology of Lizard *Eremias persica*. *Zool. Zhur. 51,* 310-312 (in Russian).

Bogdanov, O. P. and Zinyakova, M. P. Z. (1965). On the diurnal activity of *Vipera lebetina turanica* on the Nuratau Ridge. *Zool. Zhur. 44,* 1733-1734 (in Russian).

Bogert, C. M. (1940). Herpetological results of the Vernay Angola Expedition with notes on African reptiles in other collections. Part I. Snakes, including an arrangement of African Colubridae. *Bulletin of the American Museum of Natural History 77,* 1-107.

Bohme, W. (1977). Further specimens of the rare cat snake, *Telescopus rhinopoma* (Blanford 1874) (Reptilia, Serpentes, Colubridae). *Journal of Herpetology 11,* 201-205.

Bolanos, R., Flores, A., Taylor, R. T. and Cerdas, L. (1974). Color patterns and venom characteristics in *Pelamis platurus*. *Copeia* 909-911.

Boltt, R. E. and Ewer, R. F. (1964). The functional anatomy of the head of the puff adder, *Bitis arietans* (Merr.). *Journal of Morphology 114,* 83-106.

Bons, J. (1960). Effets de l'amincissement de la roque de l'oeuf sur le developpement du lezard *Acanthodactylus pardalis*. *C. R. Soc. Biol. Paris 154,* 490-492.

Bons, J. (1963). Notes sur *Blanus cinereus* (Vandelli) description d'une sous-espece Marocaine: *Blanus cinereus mettetali* subsp. nov. *Bull. Soc. Sci. nat. phys. Maroc. 43,* 95-107.

Bons, J. and Bons, N. (1960). Notes sur la reproduction et le developpement de *Chamaeleo chamaeleon* (L.). *Bull. Soc. Sci. nat. phys. Maroc 40,* 323-335.

Bons, J. and Girot, B. (1963). Revision de l'espece *Acanthodactylus scutellatus* (Lacertidae, Saurien). *Bull. Soc. Sci. nat. phys. Maroc 42,* 311-334.

Borcea, M. (1979). Variabilitat einiger metrischer und qualitativer charactere der population *Lacerta agilis agilis* Linnaeus aus der Moldau (Rumanien). *Zool. Anz. 202,* 86-98.

Borchwardt, V. G. (1977). Development of the vertebral column in embryogenesis of *Lacerta agilis*. *Zool. Zhur. 56,* 576-587 (in Russian).

Botte, V. (1973a). Morphology and histochemistry of the oviduct of the lizard, *Lacerta sicula*: The annual cycle. *Boll. Zool. 40,* 305-314.

Botte, V. (1973b). Some aspects of oviducal biochemistry in the lizard, *Lacerta sicula* in relation to the annual cycle. *Boll. Zool. 40,* 315-321.

Botte, V. (1974). The hormonal control of the oviduct in the lizard *Lacerta sicula* Raf. I. The effects of ovariectomy and steroid replacement. *Monitore zool. ital. 8,* 47-54.

Botte, V., Angelini, F. and Picariello, O. (1978). Autumn photothermal regimes and spring reproduction in the female lizard *Lacerta sicula*. *Herpetologica 34,* 298-302.

Botte, V. and Delrio, G. (1965). Ricerche istochimiche sulla distribuzione dei 3- e 17-chetosteroidi e di alcuni enzimi della steroidogenesi nell'ovario di *Lacerta sicula*. *Boll. Zool. 32,* 191-195.

Boulenger, G. A. (1905). Description of three new snakes discovered in South Arabia by Mr. C. W. Bury. *Ann. Mag. Nat. Hist. (ser. 7) 16,* 178-180.

Boulenger, G. A. (1913). On the geographical races of *Vipera ammodytes*. *Ann. Mag. Nat. Hist. (ser. 8) 11,* 283-287.

Boulenger, G. A. (1920). *Monograph of the Lacertidae I*. London, 352 pp.

Boulenger, G. A. (1921). *Monograph of the Lacertidae II.* London, 451 pp.

Bourgeois, M. (1968). Contribution a la morphologie comparee du crane des ophidiens de l'Afrique Centrale. *Publ. Univ. Off. Congo, Lubumbashi 18,* 1-293.

Branch, W. R. and Wade, E. O. Z. (1976). Hemipenial morphology of British snakes. *British Journal of Herpetology 5,* 548-553.

Brelih, S. (1961). Seven new races of the species *Lacerta (Podarcis) sicula* Rafinesque from the Rovinj-Porec Region. *Biol. Vest. 9,* 71-91.

Broadley, D. G. (1966). A review of the African Stripe-bellied sand-snakes of the genus *Psammophis. Arnoldia Rhodesia 2(36),* 1-9.

Broadley, D. G. (1968). A review of the African cobras of the genus *Naja* (Serpentes: Elapinae). *Arnoldia Rhodesia 3(29),* 1-14.

Broadley, D. G. (1974). A review of the cobras of the *Naja nigricollis* complex in southwestern Africa (Serpentes: Elapidae). *Cimbebasia (ser. A) 2,* 155-162.

Broadley, D. G. and Parker, R. H. (1976). Natural hybridization between the puff adder and gaboon viper in Zululand (Serpentes: Viperidae). *Durban Museum Novitates 11,* 77-83.

Bruck, G. (1968). Zur herpetofauna Afghanistans. *Vestnik Ceskoslovenske Spolecnosti Zoologicke Acta Societatis Zoologicae Bohemoslovacae 32,* 201-208.

Bruno, S. (1967). Sulla *Vipera ammodytes* (Linnaeus 1758) in Italia. *Mem. Mus. Civ. Stor. Nat. Verona 15,* 289-336.

Bruno, S. (1970). Anfibi e Rettili di Sicilia. *Atti della Accad. Gioenia di Scienze naturali in Catania (ser. 7) 2,* 1-144.

Bruno, S. (1975). Note riassuntive sull'Erpetofauna dell'Isola di Montecristo (Arcipelago Toscano, Mare Tirreno). *Lav. Soc. ital. Biogeogr. 5,* 1-98.

Bruno, S. (1976). L'ornamentazione della *Vipera aspis* (L., 1758) in Italia. *Atti Soc. ital. Sci. nat. Museo civ. Stor. nat. Milano 117,* 165-194.

Bruno, S. and Hotz, H. (1976). *Coluber hippocrepis* auf der Insel Sardinien (Reptilia, Serpentes, Colubridae). *Salamandra 12,* 69-86.

Buchholz, K. F. (1960). Zur Kenntnis von *Lacerta peloponnesiaca* (Reptilia: Lacertidae). *Bonn. zool. Beitr. 11,* 87-107.

Bund, C. F. van de (1964). De verspreiding van de reptilen en amphibieen in Nederland. *Lacerta 22,* 1-72.

Buresch, I. and Beskov, V. (1965). Wird die Giftschlange *Vipera aspis* L. In Bulgarien angetroffen. *Bull. Inst. Zool. Mus., Acad. Bulgare Sci. 18,* 5-30.

Burger, W. L. and Natsuno, T. (1974). A new genus for the Arafura Smooth seasnake and redefinitions of other seasnake genera. *The Snake 6,* 61-75.

Busack, S. D. (1975). Biomass estimates and thermal environment of a population of the fringe-toed lizard, *Acanthodactylus pardalis. British Journal of Herpetology 5,* 457-459.

Capocaccia, L. (1968). *Anfibi e Rettili.* Mondadori, Milan.

Casimir, M. J. (1970). Zur herpetofauna des Iran und Afghanistan. *Datz. 23,* 150-154.

Castanet, J. (1974). Marques squelettiques de croissance chez la vipere aspic. *Zool. Scripta 3,* 137-151.

Castanet, J. and Nolleau, G. (1974). Donnees experimentales sur la valeur des marques squelettiques comme indicateur de l'age chez *Vipera aspis* (L.) (Ophidia, Viperidae). *Zool. Scripta 3,* 201-208.

Cattaneo, A. (1975). Presenza di *Elaphe longissima longissima* melanica a Castelfusano (Roma). *Atti Soc. Ital. Sci. Nat. Mus. Civ. Storia Nat. Milano 116,* 251-262.

Cesari, E. and Boquet, P. (1937). Recherches sur le venin de la vipere lebetine *(Vipera lebetina). C. r. Soc. Biol. 124,* 335-337.

Cesari, E. and Boquet, P. (1939). Sur le mecanisme de la detoxication du venin de *Vipera aspis* par l'aldehyde formique. *C. r. Soc. Biol. 130,* 19-23.

Cherchi, M. A. and Spano, S. (1964). Un nuova specie di *Tropiocolotes* del Sud Arabia Spedizione Scortecci nell'Hadramut (1962). *Bollettino dei Musei e degli Istituti Biologici dell'Universita di Genova 32*, 29-34.

Cherlin, V. A. (1981). The new saw-scaled viper *Echis multisquamatus* sp. nov. from southwestern and middle Asia. *Proceedings of the Zoological Institute of the Academy of Sciences, U.S.S.R. 101*, 92-95.

Cherlin, V. A. and Tsellarius, A. J. (1981). The behaviour dependence of the saw-scaled viper, *Echis multisquamatus* Cherlin 1981 on temperature in south Turkmenia. *Proceedings of the Zoological Institute of the Academy of Sciences, U.S.S.R. 101*, 96-108.

Cheymol, J., Boquet, P., Bourillet, F., Detrait, J. and Roch-Arveiller, M. (1973). Compariason des principales proprietes pharmacologiques de differents venins d'*Echis carinatus* (Viperides). *Archs. int. Pharmacodyn. Ther. 205*, 293-304.

Childress, J. R. (1970). Observations on the reproductive cycle of *Agama stellio picea. Herpetologica 26*, 149-155.

Choubey, B. J. (1970). Seasonal changes in the weight and histology of male sex and accessory organs of the Indian garden lizard *Calotes versicolor* (Daud.). *Bhagalpur Univ. Nat. Sci. Journal 3*, 21-38.

Choubey, B. J. and Thapliyal, J. P. (1966). Agonadal garden lizard: *Calotes versicolor. Naturwissenschaften 53*, 618.

Clark, R. J. (1972). Notes on a third collection of reptiles made in Turkey. *British Journal of Herpetology 4*, 258-262.

Clark, R. J. (1973). Report on a collection of reptiles from Cyprus. *British Journal of Herpetology 5*, 357-360.

Clark, R. J. and Clark, E. D. (1970). Notes on four lizard species from the Peloponnese, Greece: *Algyroides moreoticus* Bibron and Bory, *Anguis fragilis peloponnesiacus* Stepanek, *Ophiomorus punctatissimus* (Bibron and Bory) and *Ophisaurus apodus* (Pallas). *British Journal of Herpetology 4*, 135-137.

Clark, R. J., Clark, E. D. and Anderson, S. C. (1966). Report on two small collections of reptiles from Iran. *Occasional Papers of the California Academy of Science 55*, 1-9.

Clark, R. J., Clark, E. D., Anderson, S. C. and Leviton, A. E. (1969). Report on a collection of amphibians and reptiles from Afghanistan. *Proceedings of the California Academy of Science (ser. 4) 36*, 279-316.

Cloudsley-Thompson, J. L. (1972). Site tenure and selection in the African gecko *Tarentola annularis* (Geoffroy). *British Journal of Herpetology 4*, 286-292.

Clover, R. C. (1979). Phenetic relationships among populations of *Podarcis sicula* and *P. melissellensis* (Sauria: Lacertidae) from islands in the Adriatic Sea. *Systematic Zoology 28*, 284-298.

Cogger, H. G. (1966). The status of the 'elapid' snake *Tropidechis dunensis* De Vis. *Copeia* 893.

Cooper, J. S. and Poole, D. F. G. (1973). The dentition and dental tissue of the agamid lizard, *Uromastyx. Journal of Zoology, London 169*, 85-100.

Corbett, K. F. and Tamarind, D. L. (1979). Conservation of the sand lizard, *Lacerta agilis*, by habitat management. *British Journal of Herpetology 5*, 799-823.

Corkill, N. L. (1932). The snakes of Iraq. *Journal of the Bombay Natural History Society 35*, 552-572.

Corkill, N. L. and Cochrane, J. A. (1966). The snakes of the Arabian Peninsula and Socotra. *Journal of the Bombay Natural History Society 62*, 475-506.

Crespo, E. G. (1972). Repteis de Portugal Continental das coleccoes do Museu Bocage. *Arquivos do Museu Bocage (ser. 2) 3*, 447-612.

Cruce, M. (1977). Structure et dynamique d'une population de *Lacerta t. taurica* Pallas. *La Terre et la Vie 31*, 611-636.

Cruce, M. and Leonte, A. (1973). Distributia topografica si dinamica unei populatii de *Lacerta taurica taurica* Pall. *St. si Cerc. Biol. (ser. zool.) 25,* 593-600.

Cyren. O. (1941). Beitrage zur Herpetologie der Balkan halbinsel. *Mitt. konigl. Naturw. Inst. Sofia 14,* 36-152.

Daan, S. (1967). Variation and taxonomy of the hardun, *Agama stellio* (Linnaeus 1758) (Reptilia, Agamidae). *Beaufortia 14,* 109-134.

Dalcq, A. M. (1920). Le cycle saissonier du testicule de l'Orvet. *C. R. Soc. Biol. Paris 83,* 820-821.

Dalcq, A. M. (1921). Etude de la spermatogenese chez l'Orvet (*Anguis fragilis* Linn.). *Archs Biol. (Liege) 31,* 347-452.

Danielyan, F. D. (1965). Mechanisms of reproductive isolation in some Armenian forms of rock lizards (*Lacerta saxicola* Eversmann). *Izvestia Akad, Nauk Armenian S. S. R. (Biol.) 18,* 75-80.

Danielyan, F. D. (1967). New data on the ranges of certain subspecies of rock lizards (*Lacerta saxicola* Eversmann) in Armenia. *Biol. Zhurnal Armenii, Akad. Nauk Armenian S.S.R. 20,* 99-102 (in Russian).

Danielyan, F. D. (1971). The effects of unfavorable environmental conditions on the eggs of parthenogenetic and bisexual forms of Armenian rock lizards during incubation. *Biol. Zhurnal Armenii, Akad. Nauk Armenian S.S.R. 24,* 118-119 (in Russsian)

Darevsky, I. S. (1956). On the structure and function of the nasal gland of *Malpolon monspessulanus* Herm. (Reptilia, Serpentes). *Zool. Zhurnal 35,* 312-314 (in Russian).

Darevsky, I. S. (1957). Systematic and ecological aspects of the dispersal of the lizard *Lacerta saxicola* Eversmann in Armenia. *Zool. Sbornik Akad. Nauk Armenian S.S.R. 10,* 27-57 (in Russian).

Darevsky, I. S. (1958). Natural parthenogenesis in certain suspecies of the rock lizard *Lacerta saxicola* Eversmann. *Doklady Akad. Nauk S.S.R. 122,* 730-732.

Darevsky, I. S. (1960). The population dynamics, migration and growth in *Phrynocephalus helioscopus persicus* De Fill. in the Arax River Valley (Armenia). *Bulletin of the Natural History Society of Moscow 65,* 31-38.

Darevsky, I. S. (1962). On the origin and biological role of natural parthenogenesis in a polymorphic group of Caucasian rock-lizards, *Lacerta saxicola* Eversmann. *Zool. Zhurnal 41,* 397-408 (in Russian).

Darevsky, I. S. (1966). Natural parthenogenesis in a polymorphic group of Caucasian rock lizards related to *Lacerta saxicola* Eversmann. *Journal of the Ohio Herpetological Society 5,* 115-152.

Darevsky, I. S. (1967). Caucasian rock lizards: systematics, ecology and phylogenesis of the polymorphic groups of Caucasian rock lizards of the subgenus *Archaeolacerta*. Nauka Press, Leningrad 214 pp. (in Russian).

Darevsky, I. S. (1970). Systematic status of *Rhynchocalamus melanocephalus satunini* Nik. (Serpentes, Colubridae) previously included in the genus *Oligodon. Zool. Zhurnal 49,* 1685-1690.

Darevsky, I. S. and Danielyan, F. D. (1968). Diploid and triploid progeny arising from natural mating of parthenogenetic *Lacerta armeniaca* and *L. unisexualis* with bisexual *L. saxicola valentini. Journal of Herpetology 2,* 65-69.

Darevsky, I. S. and Danielyan, F. D. (1977). *Lacerta uzzelli* ap. nov. (Sauria, Lacertidae) — a new parthenogenetic species of rock lizard from eastern Turkey. *Proceedings of the Zoological Institute of the Academy of Sciences, U.S.S.R. 74,* 55-59 (in Russian).

Darevsky, I. S. and Danielyan, F. D. (1979). A study of the degree of genetic homogeneity in the unisexual lizard *Lacerta unisexualis* Darevsky using skin graft technique. *Proceedings of the Zoological Institute of the Academy of Sciences U.S.S.R. 89,* 65-70.

Darevsky, I. S. and Eiselt, J. (1980). Neue Felseneidechsen (Reptilia: Lacertidae) aus dem Kaukasus und aus der Turkei. *Amphibia-Reptilia 1,* 29-40.

Darevsky, I. S. and Krasilnikov, E. N. (1965). Certain traits of the blood cells of triploid hybrids of the rock-lizard, *Lacerta saxicola* Eversmann. *Doklady Acad. Nauk S.S.R. 164,* 709-711 (in Russian).

Darevsky, I. S. and Kulikova, V. N. (1961). Naturliche Parthenogenese in der polymorphen Gruppe der Kaukasischen Felseidesche (*Lacerta saxicola* Eversmann). *Zool. Jb., Syst. 89,* 119-176.

Darevsky, I. S. and Kulikova, V. N. (1962). Taxonomic characters and certain pecularities of the oogenesis of hybrids between bisexual and parthenogenetic forms of *Lacerta saxicola* Eversmann. *Cytology* (Moscow) *5,* 160-170 (in Russian).

Darevsky, I. S. and Kulikova, V. N. (1964). Natural triploidy in a polymorphic group of rock-lizards, *Lacerta saxicola* Eversmann, resulting from hybridization between bisexual and parthenogenetic forms of this species. *Doklady Acad. Nauk S.S.R. 158,* 202-205 (in Russian).

Darevsky, I. S., Kupriyanova, L. A. and Bakradze, M. A. (1978). Occasional males and intersexes in parthenogenetic species of Caucasian rock lizard (genus *Lacerta*). *Copeia* 201-207.

Darevsky, I. S. and Lukina, G. P. (1977). Rock lizards of the *Lacerta saxicola* Eversmann group (Sauria, Lacertidae) collected in Turkey by Richard and Erica Clark. *Proceedings of the Zoological Institute of the Academy of Sciences, U.S.S.R. 74,* 60-63 (in Russian).

Darevsky, I. S. and Shcherbak, N. N. (1978). *Eremias andersoni,* a new lizard (Reptilia, Lacertilia, Lacertidae) from Iran. *Journal of Herpetology 12,* 13-15.

Darevsky, I. S. and Vedmederja, V. I. (1977). A new species of rock lizard *Lacerta saxicola* Eversmann group from northeastern Turkey and adjoining regions of Adjaria. *Proceedings of the Zoological Institute of the Academy of Sciences, U.S.S.R. 74,* 50-54 (in Russian).

Dastugue, G. and Joy, M. (1943). Nouvelles recherches sur la composition du sang chez *Vipera aspis.* II. Les constituants chimiques. *C. r. Soc. Phys. Biol. 67,* 61-68.

Deraniyagala, R. Y. (1958). Pseudocombat of the monitor *Varanus bengalensis. Bulletin of the National Museum of Ceylon 28,* 11-13.

Detrait, J. and Duguy, R. (1966). Variations de toxicite du venin au cours du cycle annuel chez *Vipera aspis* L. *Annals of the Pasteur Institute 111,* 93-99.

Detrait, J., Izard, Y. and Boquet, P. (1960). Relations antigeniques entra un facteur lethal de venin d'*Echis carinatus* et les neurotoxines des venins de *Naja naja* et de *Naja nigricollis. C. r. Seanc. Soc. Biol. 154,* 1163-1165.

Dmi'el R. (1967). Studies on reproduction, growth and feeding in the snake *Spalerosophis cliffordi* (Colubridae). *Copeia* 332-346.

Dmi'el, R. and Borut, A. (1972). Thermal behavior, heat exchange and metabolism in the desert snake *Spalerosophis cliffordi. Physiol. Zool. 45,* 78-94.

Dmi'el, R. and Zilber, B. (1971). Water balance in a desert snake. *Copeia* 754-755.

Domergue, Ch. A. (1959). Cle de determination des Serpents de Tunisie et Afrique du Nord. *Arch. Inst. Pasteur Tunis 36,* 163-172.

Ducker, G. V. and Rensch, B. (1973). Die visuelle Lernkapazitat von *Lacerta viridis* und *Agama agama. Z. Tierpsychol. 32,* 209-214.

Duda, P. L. and Koul, O. (1977). Ovarian cycle in high altitude lizards from Kashmir. Part II. *Scincella himalayanum* (Boulenger). *Herpetologica 33,* 427-433.

Dufaure, J. P. (1961). Le developpement de l'appareil genital du lezard vivipare. *Arch. Anat. micr. Morph. exp. 50,* 69-80.

Dufaure, J. P. (1966). Recherches descriptives et experimentales sur les modalites et facteurs du developpement de l'appareil genital chez le lezard vivipare (*Lacerta vivipara* Jacquin). *Arch. Anat. micr. Morph. exp. 55,* 437-537.

Dufaure, J. P. (1968). L'ultrastructure des cellules interstitielles du testicule adulte chez deux reptiles lacertiliens: Le lezard vivipare (*Lacerta vivipara* Jacquin) et l'orvet (*Anguis fragilis* L.). *C. R. Acad. Sci. Paris 267,* 883-885.

Dufaure, J. P. (1970). L'ultrastructure de testicule de lezard vivipare (Reptilia, Lacertilia). I. Les cellules interstitielles. *Z. Zellforsch. 109,* 33-45.

Dufaure, J. P. (1971). L'ultrastructure de testicule de lezard vivipare (Reptilia, Lacertilia). II. Les cellules de Sertoli. Etude du glycogene. *Z. Zellforsch. 115,* 565-578.

Dufaure, J. P. and Hubert, J. (1965). Origine et migration des gonocytes primordiaux chez l'embryon de lezard vivipare (*Lacerta vivipara* Jacquin). *C. R. Acad. Sci. Paris 261,* 237-240.

Duguy, R. (1962). Biologie de la latence hivernale chez *Vipera aspis* L. *Vie et Milieu 14,* 311-443.

Duguy, R. (1963). Donnees sur le cycle annuel du sang circulant chez *Anguis fragilis* L. *Bulletin of the Zoological Society of France 88,* 99-108.

Duguy, R. and Saint Girons, H. (1969). Etude morphologique des populations de *Vipera aspis* (Linnaeus 1758) dans l'ouest et le sud-ouest de la France. *Bull. Mus. natn. Hist. nat. Paris 41,* 1069-1090.

Dullemeijer, P. (1956). The functional morphology of the head of the common viper *Vipera berus* (L.). *Archs neerl. Zool. 11,* 387-497.

Dummermuth, S. (1977a). Behandlung und Folgen eines Vipernbisses *(Vipera a aspis). Aquaria 24,* 34-35.

Dummermuth, S. (1977b). Pflege und Zucht der Wurfelnatter *(Natrix t. tessellata). Aquaria 24,* 43-44.

Dunson, W. A., Dunson, M. K. and Keith, A. D. (1978). The nasal gland of the montpellier snake *Malpolon monspessulanus:* Fine structure, secretion, composition and a possible role in reduction of dermal water loss. *Journal of Experimental Zoology 203,* 461-473.

Duvdevani, I. (1972). The anatomy and histology of the nasal cavities and the nasal salt gland in four species of fringe-toed lizards, *Acanthodactylus* (Lacertidae). *Journal of Morphology 137,* 353-364.

Duvdevani, I. and Borut, A. (1974). Oxygen consumption and evaporative water loss in four species of *Acanthodactylus* (Lacertidae). *Copeia* 155-164.

Edgar, W., Warrell, M. J., Warrell, D. A. and Prentice, C. R. M. (1980). The structure of soluble fibrin complexes and fibrin degradation products after *Echis carinatus* bite. *British Journal of Haematology 44,* 471.

Eggert, B. (1935). Zur Morphologie und Physiologie der Eidechsen Schilddruse. I. Das jahrezeitliche Verhalten der Schilddruse von *Lacerta agilis. L. vivipara* und *L. muralis. Z. wiss. Zool. 147,* 205-262.

Eikhorst, W., Eikhorst, R., Nettmann, H. K. and Rykena, S. (1979). Beobachtungen an der Spanischen kieleidechse, *Algyroides marchi* Valverde 1958 (Reptilia: Sauria: Lacertidae). *Salamandra 15,* 254-263.

Eiselt, J. (1940). Der Rassenkreis *Eumeces schneideri* Daudin (Scincidae, Rept.). *Zool. Anz. 131,* 209-228.

Eiselt, J. (1961). *Catalogus Faunae Austriae. 21: Amphibia, Reptilia.* Springer-Verlag, Wien.

Eiselt, J. (1965). Bericht uber eine zoologische Sammelreise nach Sudwest-Anatolien im April/Mai 1964. *Ann. Naturhist. Mus. Wien 68,* 401-406.

Eiselt, J. (1968). Ein Beitrag zur Taxonomie der Zagros Eidechse, *Lacerta princeps* Blanf. *Ann. Naturhist. Mus. Wien 72,* 409-434.

Eiselt, J. (1969). Zweiter Beitrag zur Taxonomie der Zagroseidechse *Lacerta princeps* BLANFORD. *Ann. Naturhist. Mus. Wien 73,* 209-220.

Eiselt, J. (1979). Ergebnisse zoologischer Sammelreisen in der Turkei *Lacerta cappadocica* Werner 1902 (Lacertidae, Reptilia). *Ann. Naturhist. Mus. Wien 82,* 387-422.

Elkan, E. (1976). The micro-anatomy of the skin of *Ophisaurus apodus* Pallas (Anguidae, Sauria). *British Journal of Herpetology 5,* 529-532.

El-Toubi, M. R. (1938). The osteology of the lizard *Scincus scincus* (Linn.). *Bulletin of the Faculty of Science, Cairo University 14,* 5-38.

El-Toubi, M. R. and Bishai, H. M. (1959). On the anatomy and histology of the alimentary tract of the lizard *Uromastyx aegyptia* (Forskal). *Bulletin of the Faculty of Science, Cairo University 34,* 13-50.

Eyal-Giladi, H. (1964). The development of the chondrocranium of *Agama stellio. Acta zool. Stockholm 45,* 139-165.

Fainaru, M., Eisenberg, S., Manny, N. and Hershko, C. (1974). The natural course of defibrination syndrome caused by *Echis colorata* venom in man. *Thromb. Diath. Haemorrh. 31,* 420-428.

Fein, A., Bdolah, A. and Kochva, E. (1971). Developmental pattern of enzyme secretion in embryonic venom gland of *Vipera palaestinae* (Ophidia, Reptilia). *Devl. Biol. 24,* 520-532.

Fejervary-Langh, A. M. (1943). Beitrage und Berichtigungen zum Reptilien — Teil des ungarischen Faunenkataloges. *Fragmenta Faunistica Hungarica 6,* 81-98.

Filosa, S. (1973). Biological and cytological aspects of the ovarian cycle in *Lacerta sicula sicula* Raf. *Monitore zool. ital. (N.S.) 7,* 151-165.

Fischer, K. (1970). Untersuchungen zur Jahresperiodik der Fortpflanzung bei maennlichen Ruineneideschsen (*Lacerta sicula campestris* Betta). *Z. vergl. Physiol. 66,* 273-293.

Flower, S. S. (1933). Notes on the recent reptiles and amphibians of Egypt, with a list of the species recorded from that Kingdom. *Proceedings of the Zoological Society of London* 735-851.

Forcart, L. (1950). Amphibien und Reptilien von Iran. *Verh. Naturf. Ges. Basel 61,* 151-156.

Foxon, G. E. H., Griffith, J. and Price, M. (1956). The mode of action of the heart of the green lizard, *Lacerta viridis. Proceedings of the Zoological Society of London 126,* 145-157.

Frankenberg, E. (1974). Vocalization of males of three geographical forms of *Ptyodactylus* from Israel (Reptilia: Sauria: Gekkoninae). *Journal of Herpetology 8,* 59-70.

Frankenberg, E. (1978). Calls of male and female tree geckos, *Cyrtodactylus kotschyi. Israel Journal of Zoology 27,* 53-56.

Frenkel, G. and Kochva, E. (1970). Visceral anatomy of *Vipera palaestinae* an illustrated presentation. *Israel Journal of Zoology 19,* 145-163.

Fretey, J. (1975). *Guide des reptiles et batraciens de France.* Hatier, Paris.

Froesch-Franzon, P. (1974). Wie ein *Coluber viridiflavus* eine *Lacerta m. muralis* verschlang. *Aquaria 21,* 136-140.

Fuhn, I. E. (1956). Contributi la sistematica si ecologia gusterilor din R. P. R. I. Gusternul vargat — *Lacerta trilineata media* Lantz + Cyren. *Bull. Stiint., Sect. Biol. Stiint. Agric. 8,* 469-482.

Fuhn, I. E. (1969). Revision and redefinition of the genus *Ablepharus* Lichtenstein 1823 (Reptilia, Scincidae). *Revue Roum. Biol.-Zool. 14,* 23-41.

Gabaeva, N. S. (1970). Histogenesis of follicular epithelium and formation of vitelline membrane of *Lacerta agilis* and *Agama caucasica* oocytes. *Arkh. Anat. Gistol. Embriol. 59,* 28-39.

Gabe, M. (1969). Emplacement des cellules a gastrine dans l'estomac de quelques sauropsides et batraciens. *C. r. hebd. Seanc. Acad. Sci. Paris (ser. D) 268,* 3088-3090.

Gabe, M. (1971). Repartition des cellules histamin-ergiques dans la paroi gastrique de quelques Reptiles. *C. r. hebd. Seanc. Acad. Sci Paris (ser. D) 273,* 2287-2289.

Gabe, M. (1972). Donnees histologiques sur les cellules a gastrine des sauropsides. *Archs. hebd. Anta. microsc. Morph. exp. 61,* 175-200.

Gabe, M. and Saint Girons, H. (1962). Donnees histophysiologiques sur l'elaboration d'hormones sexuelles au cours du cycle reproducteur chez *Vipera aspis* (L.). *Aca anat. 50,* 22-51.

Gabe, M. and Saint Girons, H. (1965). Contribution a la morphologie comparee du cloaque et des glandes epidermoides de la region cloacale des lepido sauriens. *Mem. Mus. natn. Hist. nat. Paris (ser. A) 33,* 149-292.

Gabe, M. and Saint Girons, H. (1969). Donnees histologiques sur les glandes salivaires des Lepidosauriens. *Mem. Mus. natn. Hist. nat. Paris (ser. A) 58,* 1-112.

Gabe, M. and Saint Girons, H. (1972). A contribution to the histological study of the stomach in Lepidosauria (Reptilia). *Zool. Jb. Anat. 89,* 579-599.

Gabriel-Robez, O. and Clavert, J. (1980). Teratogenic and lethal properties of the various fractions of venom of the viper *Vipera aspis. Acta anat. 108,* 226.

Gans, C. (1952). The functional morphology of the egg-eating adaptions in the snake genus *Dasypeltis. Zoologica N.Y. 37,* 209-243.

Gans, C. (1959). A taxonomic revision of the African snake genus *Dasypeltis* (Reptilia, Serpentest). *Ann. Mus. R. Congo Belge. Terv. (ser. 8), Sci. Zool. 74,* 1-237.

Gans, C. (1962). Notes on amphisbaenids (Amphisbaenia, Reptilia) I. on the name *Amphisbaena reticulata* Holmer 1787. *British Journal of Herpetology 3,* 12-13.

Gans, C. (1967). A Checklist of Recent Amphisbaenians (Amphisbaenia, Reptilia). *Bulletin of the American Museum of Natural History 135,* 61-106.

Gans, C. and Latifi, M. (1973). Another case of presumptive mimicry in snakes. *Copeia* 801-802.

Gaspereti, J. (1977). Snakes in Arabia. *Journal of the Saudi Arabian Natural History Society 19,* 3-16.

Gavish, L. (1979). Conditioned-response of snakes *(Malpolon monspessulanus)* to light (Reptilia, Serpentes, Colubridae). *Journal of Herpetology 13,* 357-359.

Gerzeli, G. and Piceis Polver, P. de (1970). The lateral nasal gland of *Lacerta viridis* under different experimental conditions. *Monitore zool. ital. (N. S.) 4,* 191-200.

Gislen, T. and Kauri, H. (1959). Zoogeography of the Swedish Amphibians and Reptiles. *Acta Vertebratica 1,* 193-397.

Gitter, S., Kochwa, S., Vries, A. de and Leffkowitz, M. (1957). Electrophoretic fractions of *Vipera xanthina palaestinae* venom. *American Journal of tropical Medicine and Hygiene 6,* 180-189.

Gitter, S., Levi, G., Kochwa, S., Vries, A. de, Rechnic, J. and Casper, J. (1960). Studies on the venom of *Echis colorata. American Journal of tropical Medicine and Hygiene 9,* 391-399.

Gitter, S., Moroz-Perlmutter, C., Boss, J. H., Livni, E., Rechnic, J., Goldblum, N. and Vries, A. de (1962). Studies on snake venoms of the Near East *Walterinnesia aegyptia* and *Pseudocerastes fieldii*. *American Journal of tropical Medicine and Hygiene 11*, 861-868.

Glandt, D. (1976). Okologische beobachtungen an niederrheinischen *Lacerta*-populationen, *Lacerta agilis* and *Lacerta vivipara* (Reptilia, Sauria, Lacertidae). *Salamandra 12*, 127-139.

Glandt, D. (1977). Uber eine *Lacerta agilis/Lacerta vivipara* population, nebst bemerkungen zum sympatrieproblem. *Salamandra 13*, 13-21.

Goel, S. C. (1976). On the mechanism of water uptake by the developing eggs of *Calotes versicolor*. *Experientia 32*, 1331-1333.

Gonzalez, D. (1976). Datos sobre morfologia y biometria de *Vipera aspis* (Viperidae). *Miscelanea Zoologica 3*, 181-193.

Gouder, B. Y. M., Nadkarni, V. B. and Appaswamy Rao, M. (1979). Histological and histochemical studies on Follicular Atresia in the ovary of the lizard *Calotes versicolor* (Reptilia, Sauria, Agamidae). *Journal of Herpetology 13*, 451-456.

Graham, J. B. (1974a). Aquatic respiration in the sea snake *Pelamis platurus*. *Respir. Physiol. 21*, 1-7.

Graham, J. B. (1974b). Body temperature of the sea snake *Pelamis platurus*. *Copeia* 531-533.

Graham, J. B., Gee, J. H. and Robison, F. S. (1975). Hydrostatic and gas exchange functions of the lung of the sea snake *Pelamis platurus*. *Comparative Biochemistry and Physiology A50*, 477-482.

Graham, J. B., Rubinoff, I. and Hecht, M. K. (1971). Temperature physiology of the sea snake *Pelamis platurus*. An index of its colonization potential in the Atlantic Ocean. *Proceedings of the National Academy of Sciences 68*, 1360-1363.

Grandison, A. G. C. (1961). Preliminary notes on the taxonomy of *Tarentola annularis* and *T. ephippiata* (Sauria: Gekkonidae). *Zoologische Mededelingen 38*, 1-14.

Greene, H. W. (1973). Defensive tail display by snakes and amphisbaenians. *Journal of Herpetology 7*, 143-161.

Greer, A. E. (1970). A subfamilial classification of scincid lizards. *Bulletin of the Museum of Comparative Zoology 139*, 151-183.

Greer, A. E. (1979). A phylogenetic subdivision of Australian skinks. *Records of the Australian Museum 32*, 339-371.

Gregory, P. T. (1980). Physical factor selectivity in the fossorial lizard *Anguis fragilis*. *Journal of Herpetology 14*, 95-99.

Greschik, E. (1917). Uber den Darmkanal von *Ablepharus pannonicus* Fitz. und *Anguis fragilis* L. *Anat. Anz. 60*, 70-80.

Grossenbacher, K. and Brand, M. (1973). *Schlussel zur Bestimmung der Amphibien und Reptilien der Schweiz*. Naturhistorisches Museum, Bern.

Grotto, L., Moroz, C., Vries, A. de and Goldblum, N. (1967). Isolation of *Vipera palaestinae* hemorrhagin and distinction between its hemorrhagic and proteolytic activities. *Biochim. biophys. Acta 133*, 356-362.

Grunwald, E. (1931). La torsion intestinale chez les reptiles. *Archs. Anat. Histol. Embryol. 14*, 167-203.

Gruschwitz, M. (1978). Untersuchungen zu vorkommen und lebensweise der wurfelnatter *(Natrix t. tessellata)* im bereich der flusse Mosel und Lahn (Rheinland-Pfalz) (Reptilia: Serpentes: Colubridae). *Salamandra 14*, 80-89.

Guibe, J. (1957). Reptiles d'Iran recoltes par M. F. Petter. *Bulletin du Museum National d'Histoire Naturelle (ser. 2) 29*, 136-143.

Guibe, J. (1966a). Reptiles et Amphibiens recoltes par la Mission Franco-Iranienne. *Bulletin du Museum National d'Histoire Naturelle (ser. 2) 38*, 97-98.

Guibe, J. (1966b). Contribution a l'etude des *Microgecko* Nikolsky et *Tropiocolotes* Peters (Lacertilia, Gekkonidae). *Bulletin du Museum National d'Histoire Naturelle (ser. 2) 38,* 337-346.

Guillaume, C. P. (1976). Etude biometrique des especes *Lacerta hispanica* Stein dachner 1870 et *Lacerta muralis* Laurenti 1768. *Bull. Soc. Zool. France 101,* 489-502.

Gygax, P. (1971). Entwicklung, Bau und Kunktion der Giftdruse (Duvernoy's gland) von *Natrix tessellata. Acta tropica 28,* 225-274.

Haas, G. (1930). Uber die Kaumuskulatur und die Schadelmechanik einiger Wuhlschlangen. *Zool. Jb., Abt. Anat. 52,* 95-218.

Haas, G. (1931a). Die Kiefermuskulatur und die Schadelmechanik der Schlangen in vergleichender Darstellung. *Zool. Jb., Abt. Anat. 53,* 127-198.

Haas, G. (1931b). Uber die Morphologie der Kiefermuskulatur und die Schadel-mechanik einiger Schlangen. *Zool. Jb., Abt. Anat. 54,* 333-416.

Haas, G. (1932). Untersuchungen uber den Kieferapparat und die verwandtschaft-lichen Zusammenhange der Schlangen. *Forsch. Fortschr. 8,* 207-210.

Haas, G. (1937). The structure of the nasal cavity in *Chamaeleo chamaeleon* (Linnaeus). *Journal of Morphology 61,* 433-451.

Haas, G. (1943). On a collection of reptiles from Palestine, Transjordan and Sinai. *Copeia* 10-15.

Haas, G. (1950). A new *Atractaspis* (Mole Viper) from Palestine. *Copeia* 52-53.

Haas, G. (1951a). Remarks on the status of the lizard *Eremias olivieri* Audouin. *Copeia* 274-276.

Haas, G. (1951b). On the present state of our knowledge of the herpetofauna of Palestine. *Bulletin of the Research Council of Israel 1,* 67-95.

Haas, G. (1952a). Two collections of reptiles from Iraq with descriptions of two new forms. *Copeia* 20-22.

Haas, G. (1952b). Remarks on the origin of the herpetofauna of Palestine. *Revue de la Faculte des Sciences de l'Universite d'Istanbul (ser. B) 17,* 95-105.

Haas, G. (1957). Some amphibians and reptiles from Arabia. *Proceedings of the California Academy of Sciences (ser. 4) 29,* 47-86.

Haas, G. (1961). On a collection of Arabian reptiles. *Annals of the Carnegie Museum 36,* 19-28.

Haas, G. and Battersby, J. C. (1959). Amphibians and reptiles from Arabia. *Copeia* 196-202.

Haas, G. and Werner, Y. L. (1969). Lizards and snakes from Southwestern Asia, collected by Henry Field. *Bulletin of the Museum of Comparative Zoology 138,* 327-406.

Hager, P. K. (1905). Die Kiefermuskeln der Schlangen und ihre Beziehungen zu den Speicheldrusen. *Zool. Jb., Abt. Anat. 22,* 173-224.

Hahn, D. E. (1978a). Liste der rezenten Amphibien und Reptilien. Scolecophidia. Anomalepidae, Leptotyphlopidae and Typhlopidae. *Das Tierreich 101,* 1-65.

Hahn, D. E. (1978b). A brief review of the genus *Leptotyphlops* (Reptilia, Serpentes, Leptotyphlopidae) of Asia, with description of a new species. *Journal of Herpetology 12,* 477-489.

Halfpenny, G. and Bellairs, A. d'A. (1976). A black grass snake. *British Journal of Herpetology 5,* 541-542.

Harding, K. A. and Welch, K. R. G. (1980). *Venomous snakes of the World. A Check-list.* Pergamon Press 188 p.

Hassan, F. and El Hawary, M. F. S. (1977). Fractionation of the snake venoms of *Cerastes cerastes* and *Cerastes vipera. Toxicon 15,* 170-173.

Hecht, M. K., Kropach, C. and Hecht, B. M. (1974). Distribution of the yellow-bellied sea snake *Pelamis platurus,* and its significance in relation to the fossil record. *Herpetologica 30,* 387-395.

Hellmich, W. (1959). Bemerkungen zu einer kleinen Sammlung von Amphibien und Reptilien aus Sud-Persien. *Opusc. Zool. Munchen 35*, 1-9.

Hett, J. (1924). Das Corpus Luteum der Zauneidechse *(Lacerta agilis). Z. mikr.- anat. Forsch. 1*, 41-84.

Heusser, J. and Schlumpf, H. U. (1962). Die Todstellen bei der Barren-Ringelnatter *(Natrix natrix helvetica). Aquar. Terr. 2*, 214-218.

Hibbard, E. and LaVergne, J. (1972). Morphology at the retina of the seasnake *Pelamis platurus. Journal of Anatomy 112*, 125-136.

Hiller, U. (1977). Regeneration and degeneration of setae-bearing sensilla in scales of gekkonid lizard *Tarentola mauritanica* L. *Zool. Anz. 199*, 113-120.

Hoofien, J. H. (1957). An addition to the fauna of Sinai, *Eremias brevirostris* Blanf. (Reptilia, Lacertidae). *Ann. Mag. Nat. Hist. (ser. 12) 10*, 719-720.

Hoofien, J. H. (1962). An unusual congregation of the gekkonid lizard *Tarentola annularis* (Geoffroy). *Herpetologica 18*, 54-56.

Hoofien, J. H. (1964). Geographical variability in the common chamaeleon in Israel. *Israel Journal of Zoology 13*, 136-138.

Hoofien, J. H. (1965). On some herpetological records from Sinai and Transjordan. *Israel Journal of Zoology 14*, 122-127.

Hoofien, J. H. (1969). A note on the wall lizard of Petra, Transjordan. *Israel Journal of Zoology 18*, 39-40.

Hoofien, J. H. (1972). A taxonomic list of the reptiles of Israel and its administered areas according to the status on May 31st. 1972. *Dept. Zool. Tel-Aviv Univ.*, 1-4.

Horn, H. (1947). The embryonic development of the pituitary body in the chamaeleon. *Copeia* 262-268.

Horn, H. G. (1976). Pflege und Nachzucht von *Lacerta muralis brueggemanni. Das Aquarium M. Aqua Terra 10*, 315-317.

Howard, N. L. (1975). Phospholipase A_2 from puff adder *(Bitis arietans)* venom. *Toxicon 13*, 21-30.

Hubert, J. (1965). Premiere confirmation de l'origine extra embryonnaire des gonocytes primordiaux chez le lezard vivipare *(Lacerta vivipara* Jacquin). *C. R. Acad. Sci. Paris 261*, 4505-4508.

Hubert, J. (1966). Localisation precoce et migration des gonocytes primordiaux chez l'embryon de Vipere aspis *(Vipera aspis* L.) *C. R. Acad. Sci. Paris 262*, 1120-1122.

Hubert, J. (1967). Nouveaux caracteres cytologiques des gonocytes primordiaux de l'embryon de lezard vivipare *(Lacerta vivipara* Jacquin). *C. R. Acad. Sci. Paris 264*, 830-833.

Hubert, J. (1968a). A propos de la lignee germinale chez 2 reptiles: *Anguis fragilis* L. et *Vipera aspis* L. *C. R. Acad. Sci. Paris 266*, 231-233.

Hubert, J. (1968b). Ultrastructure des gonocytes primordiaux chez l'embryon de lezard vivipare *(Lacerta vivipara* Jacquin). *C. R. Acad. Sci. Paris 266*, 2273-2276.

Hubert, J. (1968c). Ultrastructure des gonocytes de type (amoeboide) chez l'embryon de lezard *(Lacerta vivipara* Jacquin). *C. R. Acad. Sci. Paris 267*, 1001-1003.

Hubert, J. (1970). Ultrastructure des cellules germinales au cours du developpement embryonnaire du lezard vivipare *(Lacerta vivipara* Jacquin). *Z. Zellforsch 107*, 265-283.

Hubert, J. (1971a) Etude histologique et ultrastructurale de la granulosa a certain stades de developpement du follicule ovarien chez un lezard: *Lacerta vivipara* Jacquin. *Z. Zellforsch 115*, 46-59.

Hubert, J. (1971b). Aspects ultrastructuraux des relations entre les couches folliculaires et l'ovocyte depuis la formation du follicule jusqu'au debut de la vitellogenese chez le lezard *Lacerta vivipara* Jacquin. *Z. Zellforsch 116*, 240-249.

Hubert, J. (1971c). Localisation extra-embryonnaire des gonocytes chez l'embryon d'orvet *(Anguis fragilis* L.) *Arch. Anat. microsc. Morph. exp. 60*, 261-268.

Hubert, J. (1972). Etude par la methode autoradiographique des syntheses d'ARN, d'ADN et de proteines dans les gonocytes de l'embryon de lezard vivipare (*Lacerta vivipara* Jacquin). *C. R. Acad. Sci. Paris 274*, 1383-1386.

Hubert, J. (1976). Ultrastructure des ebauches gonodiques du lezard vivipare (*Lacerta vivipara* Jacquin) pendant la periode de colonisation par les gonocytes. *Arch. Anat. microsc. Morph. exp. 65*, 241-253.

Hubert, J. and Xavier, F. (1979). Cristallisation ribosomique et hibernation chez le lezard vivipare *Lacerta vivipara* J. *C. R. Acad. Sci. Paris D288*, 635-637.

Indurkar, S. S. and Sabnis, J. H. (1976). Observations on the dietary components of the garden lizard, *Calotes versicolor* (Daudin). *Comp. physiol. Ecol. 1*, 9-12.

Ivanov, V. G. (1973). Sex chromosomes of three species of lizards of U.S.S.R. fauna. *Voprosi Gerpetologii Leningrad* 84-86 (in Russian).

Ivanov, V. G. and Fedorova, T. A. (1970). Sex heteromorphism of chromosomes in *Lacerta strigata* Eichwald. *Tsitologiya 12*, 1582-1585 (in Russian).

Ivanov, V. G. and Fedorova, T. A. (1973). Heterochromosomes in the karyotype of *Eremias arguta* Pall. *Tsitologiya 15*, 762-765 (in Russian).

Iyer, N. M. M. (1943). The habits, external features and skeletal system of *Calotes versicolor* (Daud.): II. The skull. *Journal of the Mysore University (N. S.) 4*, 115-151.

Izard, Y. and Boquet, P. (1958). Recherches sur les venins de *Vipera xanthina palaestinae* et d'*Echis carinata*. *Annals of the Pasteur Institute, Paris 94*, 583-589.

Izard, Y., Detrait, J. and Boquet, P. (1961). Variations saisonnieres de la composition du sang de *Vipera aspis*. *Annals of the Pasteur Institute, Paris 100*, 539-545.

Jackson, O. F. (1980). Effects of a bite by the sand viper *(Vipera ammodytes)*. *Lancet 2*, 686.

Jacobs, V. L. (1979). The sensory component of the facial nerve of a reptile *(Lacerta viridis)*. *Journal of Comparative Neurology 184*, 537-546.

Jacobshagen, E. (1937). Mittel und Enddarm (Rumpfdarm). In, *"Handbuch der vergleichenden Anatomie der Wirbeltiere"*, 3, 638-654. Berlin.

Janecek, J. (1976). An exceptionally large puff adder brood, *Bitis arietans*. *International Zoo Yearbook 16*, 85-86.

Jenkins, N. K. and Simkiss, K. (1968). The calcium and phosphate metabolism of reproducing reptiles with particular reference to the adder *(Vipera berus)*. *Comparative Biochemistry and Physiology 26*, 865-876.

Jeuniaux, Ch. (1963). *Chitine et Chitinolyse*. Masson, Paris.

Joly, J. and Saint Girons, H. (1975). Influence de la temperature sur la vitesse de la spermatogenese la duree de l'activite spermatogenetique et l'evolution des caracteres sexuels secondaires du lezard des murailles, *Lacerta muralis* L. (Reptilia, Lacertidae). *Arch. Anat. microsc. Morph. exp. 64*, 317-336.

Joubert, F. J. and Taljaard, N. (1978a). *Naja haje haje* (Egyptian cobra) venom. Purification, some properties and the amino acid sequences of four toxins (CM-7, CM-8, CM-9 and CM-10b) *Biochim. biophys. Acta 534*, 331-340.

Joubert, F. J. and Taljaard, N. (1978b). Purification, some properties and the primary structures of three reduced and S-carboxymethylated toxins (CM-5, CM-6 and CM-10a) from *Naja haje haje* (Egyptian cobra) venom. *Biochim. biophys. Acta 573*, 1-8.

Joubert, F. J. and Taljaard, N. (1978c). *Naja haje haje* (Egyptian cobra) venom, some properties and the complete primary structure of three toxins (CM-2, CM11 and CM12). *European Journal of Biochemistry 90*, 359-367.

Kasturirangen, L. R. (1952). The allanto-placenta of the sea snake *Hydrophis cyanocinctus* (Daudin). *Journal of the Zoological Society of India 3*, 277-290.

Khalaf, K. T. (1959). *Reptiles of Iraq, with some notes on the amphibians.* Ar-Rabitta Press, Baghdad, 96 pp.

Khalil, F. (1951). Excretion in reptiles. IV. Nitrogenous constituents of the excreta of lizards. *Journal of Biological Chemistry 189*, 443-445.

Khalil, F. and Abdel-Messeih, G. (1954). Water content of tissues of some desert reptiles and mammals. *Journal of Experimental Zoology 125*, 407-414.

Khalil, F. and Abdel-Messeih, G. (1961a). The storage of extra water by various tissues of *Uromastyx aegyptia* (Forskal). *Z. vergl. Physiol. 45*, 78-81.

Khalil, F. and Hussein, M. F. (1962). Studies on the temperature relationships of Egyptian desert reptiles, IV. On the retention of heat by *Uromastyx aegyptia, Agama pallida* and *Chalcides serpoides. Bulletin of the Zoological Society, Egypt 17*, 80-88.

Khalil, F. and Hussein, M. F. (1963). Ecological studies in the Egyptian deserts III. Daily and annual cycles of *Uromastyx aegyptia, Agama pallida* and *Chalcides serpoides* with special reference to temperature and relative humidity. *Proceedings of the Zoological Society, U. A. R. 1*, 93-108.

Khalil, F. and Yanni, M. (1959). Studies on carbohydrates in reptiles. I. Glucose in body fluids of *Uromastyx aegyptia. Z. vergl. Physiol. 42*, 192-198.

Khalil, F. and Yanni, M. (1961). Studies on carbohydrates in reptiles. III. Seasonal changes in glycogen content of tissues, and relative weights of organs of *Uromastyx aegyptia. Z. vergl. Physiol. 44*, 355-362.

Klawe, W. L. (1964). Food of the black-and-yellow sea snake, *Pelamis platurus,* from Ecuadorian coastal waters. *Copeia* 712-713.

Klemmer, K. (1957). Untersuchungen zur Osteologie und Taxionomie der europaischen Mauereidechsen. *Abhandl. Senckenb. Naturf. Ges. 496*, 1-56.

Klemmer, K. (1960). Zur Kenntnis der Gattung *Algyroides* (Reptilia: Lacertidae) auf der Iberischen Halbinsel. *Senckenberg. biol. 41*, 1-6.

Kluge, A. G. (1967). Higher taxonomic categories of gekkonid lizards and their evolution. *Bulletin of the American Museum of Natural History 135*, 1-60.

Koch, M. (1904). Demonstration einiger Geschwulste bei Tieren. *Verhandlungen der Deutschen Pathalogischen Gesellschaft 7*, 136-147.

Kochva, E. (1958). The head muscles of *Vipera palaestinae* and their relation to the venom gland. *Journal of Morphology 102*, 23-53.

Kochva, E. (1959). An extended venom gland in the Israel mole viper, *Atractaspis engaddensis* Haas 1950. *Bulletin of the Research Council of Israel 8B*, 31-34.

Kochva, E. (1960). A quantitative study of venom secretion by *Vipera palaestinae. American Journal of tropical Medicine and Hygiene 9*, 381-390.

Kochva, E. (1961). Venom secretion by *Vipera palaestinae. Teva va'Aretz 3*, 1-7 (in Hebrew).

Kochva, E. (1962). On the lateral jaw musculature of the Solenoglypha with remarks on some other snakes. *Journal of Morphology 110*, 227-284.

Kochva, E. (1963). Development of the venom gland and trigeminal muscles in *Vipera palaestinae. Acta Anat. 52*, 49-89.

Kochva, E. (1965). The development of the venom gland in the opisthoglyph snake *Telescopus fallax* with remarks on *Thamnophis sirtalis* (Colubridae, Reptilia). *Copeia* 147-154.

Kochva, E. and Gans, C. (1965). The venom gland of *Vipera palaestinae* with comments on the glands of some other viperines. *Acta anat. 62*, 365-401.

Kochva, E., Shayer-Wollberg, M. and Sobol, R. (1967). The special pattern of the venom gland in *Atractaspis* and its bearing on the taxonomic status of the genus. *Copeia* 763-772.

Kochwa, S., Izard, Y., Boquet, P. and Gitter, S. (1959). Sur la preparation d'un immunoserum equin antivenimeux au moyen des fractions neurotoxique isolees du venin de *Vipera xanthina palaestinae*. *Annals of the Pasteur Institute of Paris 97*, 370-376.

Kochwa, S., Perlmutter, C., Gitter, S., Rechnic, J. and Vries, A. de (1960). Studies on *Vipera palaestinae* venom. Fractionation by ion-exchange chromatography. *American Journal of tropical Medicine and Hygiene 9*, 374-380.

Kopeyan, C., Rietschoten, J. von, Martinez, G., Rochat, H., Miranda, F. and Lissitzky, S. (1973). Characterisation of five neurotoxins isolated from the venoms of two Elapidae snakes *Naja haje* and *Naja nigricollis*. *European Journal of Biochemistry 35*, 244-250.

Kopstein, F. and Wettstein, O. (1920). Reptilien und Amphibien aus Albanien. *Verh. zool.-bot. Ges. Wien 70*, 387-409.

Kornalik, E. (1963). Uber den Einfluss von *Echis carinatus* — toxin auf die Blutgerinnung. *Folia haematol. 80*, 73-78.

Kornalik, E. and Taborska, E. (1972). Instraspecies individual variability in the composition of *Echis carinatus* venom. *Toxicon 10*, 529.

Korneva, L. G. (1973). Morphology of female reproductive system in *Vipera lebetina* with special reference to its activity. *Zoological Journal* (Moscow) *52*, 85-93 (in Russian).

Kostanecki, K. (1926). Le caecum des vertebres. *Bull. int. Acad. pol. Sci. Lett. (ser. B) Suppl. 1*, 1-295.

Kothari, R. M. and Patil, S. F. (1975). Effect of gamma irradiation on the common garden lizard, *Calotes versicolor* (Daud.). *Journal of Herpetology 9*, 103-105.

Kozak, M. and Simecek, J. (1977). Some knowledge on keeping *Coronella austriaca* in captivity. *Fauna Bohemiae Septentrionalis 2*, 65-68 (in Czech.).

Kral, B. (1969). Notes on the herpetofauna of certain provinces of Afghanistan. *Zoologiske Listy 18*, 55-66.

Kramer, E. (1961). Variation, sexualdimorphismus, Wachstum und Taxionomie von *Vipera ursinii* (Bonaparte 1835) und *Vipera kaznakovi* Nikolsky 1909. *Revue suisse Zoologique 68*, 627-725.

Kramer, E. (1970). Revalidierte und neue Rassen der europaischen Schlangenfauna. *Lav. Soc. ital. Biogeogr. 1*, 667-676.

Krapp, F. and Bohme, W. (1978). *Natrix tessellata* in der voreifel (Reptilia: Serpentes: Colubridae: Natricínae). *Salamandra 14*, 157-159.

Kropach, C. (1971). Sea snake *(Pelamis platurus)* aggregations on slicks in Panama. *Herpetologica 27*, 131-135.

Kropach, C. (1972). *Pelamis platurus* as a potential colonizer of the Caribbean Sea. *Bulletin of the Biological Society of Washington 2*, 267-269.

Kropach, C. (1975). The yellow-bellied sea snake, *Pelamis*, in the Eastern Pacific. In Dunson, W. A. (editor), *Biology of Sea Snakes*. University Park Press, 185-213.

Kuhn, W. (1977). Waarnemingen aan de stekelstaarthagedis *Lacerta echinata*, een boombewonende groene *Lacerta*-soort. *Lacerta* (The Hague) *35*, 43.

Kuhnel, W. and Krisch, B. (1974). On the sexual segment of the kidney in the snake *(Natrix natrix)*. *Cell Tissue Research 148*, 417-429.

Labib, R. S., Halim, H. Y. and Farag, N. W. (1979). Fractionation of *Cerastes cerastes* and *Cerastes vipera* snake venoms by gel filtration and identification of some enzymatic and biological activities. *Toxicon 17*, 337-345.

Lac, J. (1968). *Obojzivelniky ad Plazy*. Stavovce Slovenska I, Bratislava.

Lampe, E. (1911). Erster Nachtrag zum Katalog der Reptilien — und Amphibien — Sammlung des Naturhistorischen Museums der Stadt Wiesbaden. *Jahrb. nassau Ver. Naturk.* (Wiesbaden) *64*, 137-236.

Landmann, L. (1979). Keratin formation and barrier mechanisms in the epidermis of *Natrix natrix* (Reptilia: Serpentes): an ultrastructural study. *Journal of Morphology 162*, 93-126.

Langerwerf, B. A. W. A. (1977). De hardoen *Agama stellio* in het terrarium. *Lacerta 35*, 84-86.

Langerwerf, B. A. W. A. (1980a). The successful breeding of lizards from temperate regions. In, *The care and breeding of captive reptiles*. British Herpetological Society 37-46.

Langerwerf, B. A. W. A. (1980b). The Caucasian greenlizard, *Lacerta strigata* Eichwald 1831, with notes on its reproduction in captivity. *British Herpetological Society Bulletin 1*, 23-26.

Langerwert, B. (1975). Reptielen en amfibieen in Finland. *Lacerta 34*, 4-7.

Lantz, L. A. and Cyren, O. (1936). Contribution a la connaissance de *Lacerta saxicola* Eversmann. *Bulletin de societe zoologique de France 61*, 159-181.

Lantz, L. A. and Cyren, O. (1939). Contribution a la connaissance de *Lacerta brandtii* De Filippi et de *Lacerta parva* Boulenger. *Bulletin de societe zoologique de France 64*, 228-243.

Lantz, L. A. and Suchow, G. F. (1934). *Apathya cappadocica urmiana* susp. nov., eine neue Eidechsenform aus dem persischen Kurdistan. *Zool. Anz. 106*, 294.

Lanza, B. (1954). Notizie su alcune lucertole italiane e descrizione di una nuova razza insulare del golfo di Salerno. *Boll. di Zool. (Torino) 21*, 133-143.

Lanza, B. (1956). Contributo alla migliore conoscenza di alcune forme italiane di *Lacerta muralis* (Laurenti) e descrizione di una nuova razza dell'Arcipelago Toscano. *Monitore zool. ital. 63*, 259-284.

Lanza, B. (1966). Su due nuove razze insulari di *Lacerta sicula* e di *Lacerta tiliguerta*. *Archo zool. ital. 51*, 511-522.

Lanza, B. (1978). On some new or interesting East African Amphibians and Reptiles. *Monitore zool. ital. (N.S.) 10*, 229-297.

Lanza, B. and Borri, M. (1969). Su alcune populazioni di *Lacerta sicula* Rafinesque dell' Archipelago Toscano. *Annali Mus. civ. Stor. nat. Giacomo Doria 77*, 671-693.

Latifi, M., Hoge, A. R. and Eliazan, M. (1966). The poisonous snakes of Iran. *Memoires Institute Butantan 33*, 735-744.

Laurent, R. F. (1950). Revision du genre *Atractaspis* A. Smith. *Memoires Inst. R. Sci. nat. Belgique 38*, 1-49.

Laurent, R. F. (1968). A re-examination of the snake genus *Lycophidion* Dumeril and Bibron. *Bulletin of the Museum of Comparative Zoology 136*, 461-482.

Lee, C. Y., Chen, Y. M. and Mebs, D. (1976). Chromatographic seperation of the venom of Egyptian black snake *(Walterinnesia aegyptia)* and pharmacological characterization of its components. *Toxicon 14*, 275-281.

Lefrou, G. and Martignoles, I. (1954). Contribution a l'etude des properties du venin d'un viperide africain: *Echis carinatus*. *Annals of the Pasteur Institute of Paris 86*, 446-457.

Lehrs, P. (1910). Ueber eine *Lacerta* aus dem hohen Libanon (*L. fraasii* n. sp.) und andere Montanformen unter den Eidechsen. *Festschrift zum sechzigsten Geburtstage Richard Hertwigs Jena 2*, 227-238.

Leviton, A. E. (1959). Report on a collection of reptiles from Afghanistan. *Proceedings of the California Academy of Sciences (ser. 4) 29*, 445-463.

Leviton, A. E. (1977). A new Lytorhynchid snake. *Journal of the Saudi Arabian Natural History Society 19*, 16-25.

Leviton, A. E. and Anderson, S. C. (1961). Further remarks on the amphibians and reptiles of Afghanistan. *Wasmann Journal of Biology 19*, 269-276.

Leviton, A. E. and Anderson, S. C. (1963). Third contribution to the herpetology of Afghanistan. *Proceedings of the California Academy of Sciences 31*, 329-339.

Leviton, A. E. and Anderson, S. C. (1967). Survey of the reptiles of the Sheikdom of Abu Dhabi, Arabian Peninsula. Part II. Systematic account of the collection of reptiles made in the Sheikdom of Abu Dhabi by John Gasperetti. *Proceedings of the California Academy of Sciences 35*, 157-192.

Leviton, A. E. and Anderson, S. C. (1970a). Review of the snakes of the genus *Lytorhynchus*. *Proceedings of the California Academy of Sciences (ser. 4) 37*, 249-279.

Leviton, A. E. and Anderson, S. C. (1970b). The amphibians and reptiles of Afghanistan, a checklist and key to the herpetofauna. *Proceedings of the California Academy of Sciences (ser. 4) 38*, 163-206.

Leviton, A. E. and Anderson, S. C. (1972). Description of a new species of *Tropiocolotes* (Reptilia: Gekkonidae) with a revised key to the genus. *Occasional Papers of the California Academy of Sciences 96*, 1-7.

Licht, P., Hoyer, H. E. and Oordt, P. G. W. J. van (1969). Influence of photoperiod and temperature on testicular recrudescence and body growth in lizards *Lacerta sicula* and *Lacerta muralis*. *Journal of Zoology, London 157*, 469-501.

List, J. C. (1966). Comparative osteology of the snake families Typhlopidae and Leptotyphlopidae. *Illinois Biological Monographs 36*, 1-112.

Liu, C. S., Huber, G. S., Lind, C. S. and Blackwell, R. Q. (1973). Fractionation of toxins from *Hydrophis cyanocinctus* venom and determination of amino acid composition and end groups of Hydrophitoxin a. *Toxicon 11*, 73-79.

Lopez, A. and Bons, J. (1981). Studies on a case of papillomatosis in the eyed lizard *Lacerta lepida*. *British Journal of Herpetology 6*, 123-125.

Lotze, H. U. (1975). Zum paarungsverhalten der askulapnatter, *Elaphe longissima*. *Salamandra 11*, 67-76.

Loveridge, A. (1941). Certain Afro-American geckos of the genus *Hemidactylus*. *Copeia* 245-248.

Loveridge, A. (1947). Revision of the African lizards of the family Gekkonidae. *Bulletin of the Museum of Comparative Zoology 98*, 1-469.

Luppa, H. (1961). Histologie, Histogenese und Topochemie der Drusen des Sauropsidenmagens I. Reptilia. *Acta histochem. 12*, 137-179.

Mahendra, B. C. (1935a). On the peculiar apertures in the vertebral centra of *Hemidactylus flaviviridis* Ruppel. *Current Science Bangalore 4*, 34.

Mahendra, B. C. (1935b). Sexual dimorphism in the Indian House-gecko *Hemidactylus flaviviridis* Ruppel. *Current Science Bangalore 4*, 178-179.

Mahendra, B. C. (1936). Contributions to the Bionomics, Anatomy, Reproduction and Development of the Indian House-gecko *Hemidactylus flaviviridis* Ruppel. Part I. *Proceedings of the Indian Academy of Sciences 4*, 250-281.

Mahendra, B. C. (1941). Contributions to the Bionomics, Anatomy, Reproduction and Development of the Indian House-gecko *Hemidactylus flaviviridis* Ruppel. Part II. The problem of Locomotion. *Proceedings of the Indian Academy of Sciences 13*, 288-306.

Mahendra, B. C. (1942). Contributions to the Bionomics, Anatomy, Reproduction and Development of the Indian House-gecko *Hemidactylus flaviviridis* Ruppel. Part III. The heart and venous system. *Proceedings of the Indian Academy of Sciences 15*, 231-252.

Maher, M. J. (1961). The effect of environmental temperature on metabolic response to thyroxine in the lizard *Lacerta muralis*. *American Zoologist 1*, 461.

Mamonov, G. (1977). Case report of envenomation by the mountain racer *Coluber ravergieri* in USSR. *Snake 9*, 27-28.

Marian, M. (1963). Einige Daten zur Fortpflanzungsbiologie der Kreuzotter (*Vipera b. berus* L.) *Vertebr. Hung. 5*, 55-67.

Marin, G. and Sabbadin, A. (1959). Sviluppo e differenziamento delle gonadi in *Lacerta sicula campestris*. *Atti Acad. naz. Lincei Rc. Sed. Solen. 26*, 59-62.

Marshall, A. J. and Woolf, F. M. (1957). Seasonal lipid changes in the sexual elements of a male snake *Vipera berus*. *Q. Microsc. Sci. 98*, 89-100.

Marx, C. and Kayser, C. (1949). Le rythme nycthermal de l'activitie chez le lezard *(Lacerta agilis, Lacerta muralis). C. r. Soc. Biol. 143*, 1375-1377.

Marx, H. (1953). The elapid genus of snakes *Walterinnesia. Fieldiana Zoology 34*, 189-196.

Marx, H. (1959). Review of the colubrid snake genus *Spalerosophis. Fieldiana Zoology 39*, 347-361.

Marx, H. (1968). Checklist of the reptiles and amphibians of Egypt. *Special Publication of U.S. Naval Medical Research Unit 3, Egypt* 1-91.

Marx, H. and Rabb, G. B. (1965). Relationships and zoogeography of the Viperine snakes (Family Viperidae). *Fieldiana Zoology 44*, 161-206.

Mathur, J. K. and Goel, S. C. (1974). A note on a tailless embryo of the lizard *Calotes versicolor. British Journal of Herpetology 5*, 420-422.

Mattison, C. and Smith, N. D. (1978). Notes on some amphibians and reptiles from Spain. *British Journal of Herpetology 5*, 775-781.

McDowell, S. B. (1967). Osteology of the Typhlopidae and Leptotyphlopidae: a critical review. *Copeia* 686-692.

McDowell, S. B. (1974). A catalogue of the snakes of New Guinea and the Solomons, with special reference to those in the Bernice P. Bishop Museum. Part I. Scolecophidia. *Journal of Herpetology 8*, 1-57.

McDowell, S. B. (1975). A catalogue of the snakes of New Guinea and the Solomons, with special reference to those in the Bernic P. Bishop Museum. Part II. Anilioidea and Pythoninae. *Journal of Herpetology 9*, 1-80.

McDowell, S. B. (1979). A catalogue of the snakes of New Guinea and the Solomons, with special reference to those in the Bernice P. Bishop Museum. Part III. Boinae and Acrochordoidea. *Journal of Herpetology 13*, 1-92.

Mendelssohn, H. (1963). On the biology of the venomous snakes of Israel. Part I. *Israel Journal of Zoology 12*, 143-170.

Mendelssohn, H. (1965). On the biology of the venomous snakes of Israel. Part II. *Israel Journal of Zoology 14*, 185-212.

Mertens, R. (1920). Uber die geographiscen Formen von *Eumeces schneideri* Daudin. *Senckenbergiana 2*, 176-179.

Mertens, R. (1924a). Ein neuer Gecko aus Mesopotamia. *Senckenbergiana 6*, 84.

Mertens, R. (1924b). Herpetologische Mitteilungen. V. Zweiter Beitrag zur Kenntnis der geographischen Formen von *Eumeces schneideri* Daudin. *Senckenbergiana 6*, 182-184.

Mertens, R. (1932). Zur Verbreitung und Systematik einiger *Lacerta* — Formen der Apenninischen Halbinsel und der Tyrrhenischen Inselwelt. *Senckenbergiana 14*, 235-259.

Mertens, R. (1937). Neues uber die Eidechsen-Fauna Istriens. *Zool. Anz. 119*, 332-336.

Mertens, R. (1946). Dritte Mitteilung uber die Rassen der Glattechse *Eumeces schneideri. Senckenbergiana 27*, 53-62.

Mertens, R. (1947). Studien zur Eldonomie und Taxonomie der Ringlenatter *(Natrix natrix). Abh. Senckenb. naturforsch. Ges. 476*, 1-38.

Mertens, R. (1952a). Amphibien und Reptilien aus der Turkei. *Revue de la Faculte des Sciences de l'Universite d'Istanbul B17*, 54-55.

Mertens, R. (1952b). Nachtrag zu "Amphibien und Reptilien aus der Turkei." *Revue de la Faculte des Sciences de l'Universite d'Istanbul B17*, 353-355.

Mertens, R. (1955). Der typus von *Vipera lebetina schweizeri*. *Senckenberg biol. 36*, 297-299.

Mertens, R. (1956). Amphibien und Reptilien aus SO-Iran 1954. *Jahreshefte des Vereins fur vaterlandische Naturkunde in Wurttemberg 111*, 90-97.

Mertens, R. (1959). Zur Kenntnis der Lacerten auf der Insel Rhodos. *Senckenberg biol. 40*, 15-24.

Mertens, R. (1961). Die Amphibien und Reptilien der Insel Korfu. *Senckenberg biol. 42*, 1-29.

Mertens, R. (1963). Liste der rezenten Amphibien und Reptilien. Helodermatidae, Varanidae, Lanthanotidae. *Das Tierreich 79*, 1-26.

Mertens, R. (1965). Bemerkungen uber einige Eidechsen aus Afghanistan. *Senckenberg. biol. 46*, 1-4.

Mertens, R. (1966a). Liste der rezenten Amphibien und Reptilien. Chamaeleonidae. *Das Tierreich 83*, 1-37.

Mertens, R. (1966b). Uber die sibirische ringelnatter, *Natrix natrix scutata*. *Abhandlungen Senckenbergischen naturforschenden Gesellschaft 47*, 117-119.

Mertens, R. (1967). Uber *Lachesis libanotica* und den Status von *Vipera bornmuelleri*. *Senckenberg. biol. 48*, 153-159.

Mertens, R. and Wermuth, H. (1960). *Die Amphibien und Reptilien Europas*. V. W. Kramer, Frankfurt 264 pp.

Minton, S. A. (1962). An Annotated key to the amphibians and reptiles of Sind and Las Bela, West Pakistan. *American Museum Novitates 2081*, 1-60.

Minton, S. A., Anderson, S. C. and Anderson, J. A. (1970). Remarks on some geckos from Southwest Asia, with descriptions of three new forms and a key to the genus *Tropiocolotes*. *Proceedings of the California Academy of Sciences (ser. 4) 37*, 333-362.

Minton, S. A. and Salanitro, S. K. (1972). Serological relationships among some colubrid snakes. *Copeia* 246-252.

Miranda, F., Kupeyan, C., Rochat, H., Rochat, C. and Lissitzky, S. (1970). Purification of animal neurotoxins. Isolation and characterization of four neurotoxins from two different sources of *Naja haje* venom. *European Journal of Biochemistry 17*, 477-484.

Mishima, S. and Okonogi, T. (1962). Studies on the effect of sea snake venom. 2. On the toxicity of *Hydrophis cyanocinctus* Daudin venom. *Jap. J. Sanit. Zool. 13*, 153.

Mittleman, M. B. (1947). Geographic variation in the sea snake *Hydrophis ornatus* (Gray). *Proceedings of the Biological Society of Washington 60*, 1-8.

Moav, B., Moroz, C. and Vries, A. de (1963). Activation of the fibrinolytic system of the guinea pig following inoculation of *Echis colorata* venom. *Toxicon 1*, 109-112.

Moffat, L. A. and Bellairs, A. d'A. (1964). The regenerative capacity of the tail in embryonic and post-natal lizards (*Lacerta vivipara* Jacquin). *J. Embryol. exp. Morph. 12*, 769-786.

Mohamed, A. H., El-Serougi, M. and Khaled, L. Z. (1969). Effects of *Cerastes cerastes* venom on blood coagulation mechanisms. *Toxicon 7*, 181.

Mohamed, A. H. and Khaled, L. Z. (1966). Effect of the venom of *Cerastes cerastes* on nerve tissue and skeletal muscle. *Toxicon 3*, 223.

Mohamed, A. H., Saleh, A. M., Ahmed, S. and Beshir, S. R. (1975). Histopathological and histochemical effects of *Naja haje* venom on kidney tissue of mice. *Toxicon 13*, 409-413.

Morita, T., Iwanaga, S. and Suzuki, T. (1976). Activation of bovine prothrombin by an activator isolated from *Echis carinatus* venom. *Thromb. Res. 8*, 59-65.

Moroz, C., Goldblum, N. and Vries, A. de (1963). Preparation of *Vipera palaestinae* antineurotoxin using carboxy-methyl-cellulose-bound neurotoxin as antigen. *Nature*, London *200*, 697-698.

Moroz, C., Goldblum, N. and Vries, A. de (1965). Biochemical and antigenic properties of a purified neurotoxin of *Vipera palaestinae* venom. *Journal of Immunology 94*, 164-171.

Moroz, C., Hahn, J. and Vries A. de (1971). Neutralization of *Vipera palaestinae* hemorrhagin by antibody fragments. *Toxicon 9*, 57-62.

Moroz, C., Vries, A. de and Sela, M. (1966). Isolation and characterisation of a neurotoxin from *Vipera palaestinae* venom. *Biochim. biophys. Acta 124*, 136-146.

Mulherkar, L. (1962). Studies on the absorption of water by the egss of the garden lizard *Calotes versicolor* (Daudin) using Trypan Blue. *Proceedings of the National Institute of Science, India B28*, 94-99.

Muller, L. (1933). Uber die erste Nachzucht der Milosringelnatter (*Natrix natrix schweizeri* L. Muller). *Mitt. Isis, Munchen* 10-20.

Muller, L. and Wettstein, O. (1933). Amphibien und Reptilien vom Libanon. *Sitzungsber. Akad. Wiss. Wien, Math.-Naturw. Kl., Abt. 1 142*, 135-144.

Muthukkaruppan, V., Kanakambika, P., Manickavel, V. and Veeraraghavan, K. V. (1970). Analysis of the development of the lizard *Calotes versicolor*. I. A series of normal stages in the embryonic development. *Journal of Morphology 130*, 479-490.

Naulleau, G. (1965). La biologie et le comportement predateur de *Vipera aspis* au Laboratoire et dans la nature. *Bull. Biol. Fr. et Belg. 99*, 395-524.

Naulleau, G. (1966). Etude complementaire de l'activite de *Vipera aspis* dans la nature. *Vie et Milieu 17*, 461-509.

Naulleau, G. (1967). Le comportement de predation chez *Vipera aspis*. *Rev. Comport. Animal 2*, 41-96.

Naulleau, G. (1968). La vipere aspic et la captivite. *Aquarama 2*, 34-35.

Naulleau, G. (1970). La reproduction de *Vipera aspis* en captivite dans des conditions artificielles. *Journal of Herpetology 4*, 113-121.

Naulleau, G. (1971). Fertility of female *Vipera aspis* as a function of the periods of mating in captivity. *Herpetologica 27*, 385-389.

Naulleau, G. (1973a). Reproduction twice in one year in a captive viper *(Vipera aspis)*. *British Journal of Herpetology 5*, 353-357.

Naulleau, G. (1973b). Contribution a l'etude d'une population melanique 'de *Vipera aspis* dans les Alpes suisses. *Bull. Soc. Sci. nat. ouest Fr. 71*, 15-20.

Naulleau, G. (1973c). Rearing the Asp Viper *(Vipera aspis)* in captivity. *International Zoo Yearbook 13*, 108-111.

Naulleau, G. (1976). La thermoregulation chez la Vipere aspic *(Vipera aspis)* etudiee par biotelemetrie dans differentes conditions artificielles experimentales. *Bull. Soc. Zool. France 101*, 726-728.

Naulleau, G. (1979). Etude biotelemetrique de la thermoregulation chez *Vipera aspis* (L.) (Reptilia, Serpentes, Viperidae). Elevee en conditions artificielles. *Journal of Herpetology 13*, 203-208.

Nedjalkov, S. and Nashkova, O. (1961). Electrophoretische und chromatographische Untersuchungen des Giftes der bulgarischen Kreuzotter *(V. ammodytes ammodytes)*. *Zbl. Bakt. 182*, 261-267.

Nikolsky, A. M. (1903). On three new species of reptiles collected by Mr. N. Zarudny in eastern Persia in 1901. *Annuaire du Musee Zoologique de L'Academie Imperiale des Science de St. Petersburg 8*, 95-98.

Nilson, G. (1976). The reproductive cycle of *Vipera berus* in SW Sweden. *Norwegian Journal of Zoology 24*, 233-234.

Nilson, G. (1980). Male reproductive cycle of the European Adder, *Vipera berus*, and its relation to annual activity periods. *Copeia* 729-737.

Nilson, G. and Andren, C. (1978). A new species of *Ophiomorus* (Sauria: Scincidae) from Kavir Desert, Iran. *Copeia* 559-564.

Nilson, G. and Andren, C. (1981a). Morphology and taxonomic status of the grass snake, *Natrix natrix* (L.) (Reptilia, Squamata, Colubridae) on the island of Gotland, Sweden. *Zoological Journal of the Linnean Society 72*, 355-368.

Nilson, G. and Andren, C. (1981b). Reply to R. S. Thorpe's comment on the Gotland grass snake. *Zoological Journal of the Linnean Society 72*, 371-372.

Nilson, G. and Sundberg, P. (1981). The taxonomic status of the *Vipera xanthina* complex. *Journal of Herpetology 15*, 379-381.

Olexa, A. (1975). Einiges zur Okologie, Ernahrung und Eiablage von *Elaphe dione* (Pallas, 1773). *Das Aquarium M. Aqua Terra 9*, 358-361.

Orlov, N. L. (1981). About the eastern range limit of the gecko *Gymnodactylus longipes* Nikolsky, 1896. *Proceedings of the Zoological Institute of the Academy of Sciences, U.S.S.R. 101*, 89-91 (in Russian).

Otis, V. (1973). Haemocytological and serum chemistry parameters of the African puff adder, *Bitis arietans*. *Herpetologica 29*, 110-116.

Ouboter, P. E. (1976). Een kruising tussen *Lacerta pityusensis kameriana* Mertens 1927 en *Lacerta pityusensis vedrae* L. Muller 1927. *Lacerta (The Hague) 34*, 138-141.

Ovadia, M., Kochva, E. and Moav, B. (1977). The neutralization mechanism of *Vipera palaestinae* neurotoxin by a purified factor from homologous serum. *Biochim. biophys. Acta 491*, 370-386.

Pandha, S. K. and Thapliyal, J. P. (1964a). Hypophysectomy in the Indian garden lizard *Calotes versicolor*. *Naturwiss 51*, 201.

Pandha, S. K. and Thapliyal, J. P. (1964b). Effects of male hormone on the renal sex segment of castrated males of the lizard *Calotes versicolor*. *Copeia* 579-581.

Pandha, S. K. and Thapliyal, J. P. (1967). Egg laying and development in the garden lizard *Calotes versicolor*. *Copeia* 121-125.

Parker, H. W. (1930). Three new reptiles from southern Arabia. *Ann. Mag. Nat. Hist. (ser. 10) 6*, 594-595.

Parker, H. W. (1931). Some reptiles and amphibians from SE Arabia. *Ann. Mag. Nat. Hist. (ser. 10) 8*, 514-522.

Parker, H. W. (1938). Reptiles and amphibians from the souther Hejaz. *Ann. Mag. Nat. Hist. (ser. 11) 1*, 481-492.

Parker, H. W. (1942). The lizards of British Somaliland. *Bulletin of the Museum of Comparative Zoology 91*, 1-101.

Parker, H. W. (1949). The snakes of Somaliland and the Sokotra Islands. *Zool. Verh. Leiden 6*, 1-115.

Parsons, T. S. and Cameron, J. E. (1977). Internal relief of the digestive tract. In, Gans, C. and Parsons, T. S. (editors), *Biology of the Reptilia*. Academic Press *6*, 159-223.

Pasteur, G. (1960). Redecouverte et validite probable du Gekkonide *Tropiocolotes nattereri* Steindachner. *Comptes Rendus des Seances Mensuelles, Societe de Sciences Naturelles et Physique du Maroc 8*, 143-144.

Pasteur, G. (1967). Un serpent endemique de Maghreb: *Sphalerosophis dolichospilus* (Werner), Colubridae. *Bull. Mus. nat. Hist. nat. Paris (ser. 2) 39*, 444-451.

Pasteur, G. (1978). Notes sur les Sauriens du genre *Chalcides* III. Description de *Chalcides levitoni* n. sp. d'Arabie saoudite (Reptilia, Lacertilia, Scincidae). *Journal of Herpetology 12*, 371-372.

Pasteur, G. (1981). A survey of the species group of the Old World Scincid genus *Chalcides. Journal of Herpetology 15*, 1-16.

Pasteur, G. and Bons, J. (1960). Catalogue des reptiles actuels du Maroc. Revision de formes d'Afrique, d'Europe et d'Asia. *Trav. Inst. Sci. cherif. (ser. zool.) 21*, 1-132.

Pasteur, G. and Girot, B. (1960). Les tarentes de l'Ouest Africain. II. *Tarentola mauritanica. Bull. Soc. Sci. nat. phys. Maroc 40*, 309-322.

Pearson, A. D. and Tamarind, D. L. (1973). Acarine parasites on the lizard *Lacerta vivipara* Jacquin. *British Journal of Herpetology 5*, 352-353.

Pellegrin, J. (1927). Les reptiles et les batraciens de l'Afrique du Nord francaise. *Compte Rendu Assoc. Franc. Avanc. Sci. Constantine 51*, 260-264.

Pernkopf, E. and Lehner, J. (1937). Vergleichende Beschreibungen des Vorderdarms bei den einzelnen Klassen der Cranioten. In, *Handbuch der vergleichenden Anatomie der Wirbeltiere*. Berlin and Vienna. *3*, 349-476.

Perschmann, C. (1956). Uber die Bedeutung der Nierenpfortaderr, insbesondere fur die Ausscheidung von Harnstoff und Harnsaure bei *Testudo hermanni* Gml. und *Lacerta viridis* Laur., sowie uber die Funktion der Harnblase bei *Lacerta viridis* Laur., *Zool. Beitr. 2*, 447-480.

Petack, R. (1975). *Eirenis modestus* (Martin 1838). *Aquar. Terr. 22*, 354-355.

Peters, G. (1962a). Die Zwergeidechse (*Lacerta parva* Boulenger) und ihre Verwandtschaftsbeziehungen zu anderen Lacertiden, insbesondere zur Libanon-Eidechse (*L. fraasii* Lehrs). *Zool. Jb. Syst. 89*, 407-478.

Peters, G. (1962b). Studien zur taxonomie Verbreitung und Oekologie der Smaragdeidechsen I. *Lacerta trilineata, viridis* und *strigata* als selbstaendige Arten. *Mitt. Zool. Mus. Berlin 38*, 127-152.

Peters, G. (1964b). Studien zur taxonomie, Verbreitung und Oekologie der Smaragdeidechsen 3. Die orientalischen populationen von *Lacerta trilineata. Mitt. Zool. Mus. Berlin 40*, 186-250.

Petter-Rousseaux, A. (1953). Recherches sur la croissance et le cycle d'activite testiculaire de *Natrix natrix helvetica* (Lacepede). *Terre et Vie 4*, 175-223.

Petzold, H. G. (1976). Eine albinotische vierstreifennatter *Elaphe quaturolineata suromates* aus Bulgarien. *Salamandra 11*, 113-118.

Petzold, H. G. (1978). Nigrinos von *Lacerta vivipara* aus der umbegung Berlins (Reptilia: Sauria: Lacertidae). *Salamandra 14*, 98-100.

Peyer, B. (1912). Die Entwicklung des Schadelskeletes von *Vipera aspis. Morph. Jahrb. 44*, 563-621.

Phelps, T. E. (1978). Seasonal movement of the snakes *Coronella austriaca, Vipera berus* and *Natrix natrix* in southern England. *British Journal of Herpetology 5*, 775-781.

Pickwell, G. V. (1971). Knotting and coiling behavior in the pelagic sea snake *Pelamis platurus* (L.). *Copeia* 348-350.

Plehn, M. (1911). Uber Geschwulste bei niederen Wirbeltieren. *2e Conference Internationale Etude Cancer* 221-242.

Poguda, A. A. (1972). The stability of the solutions of the snake toxins from *Vipera lebetina, Echis carinatus* and *Naja oxiana* venom during storage. *Biol. Nauk. 15*, 52-55.

Polozhikhina, V. F. (1965). Morphological peculiarity of the lizard *Lacerta derjugini* Nik. *Referati hauchnich soobstevenij Izdavie Moskovskogo universiteta* 95-96 (in Russian).

Pough, F. H. (1971). Adaptations for undersand respiration in the Asian sand skink *Ophiomorus tridactylus. Journal of Herpetology 5*, 72-74.

Pozzi, A. (1966). Geonomia e catalogo ragionato degli anfibi e dei rettili della Jugoslavia. *Natura 57*, 5-55.

Prestt, I. (1971). An ecological study of the viper, *Vipera berus*, in southern Britain. *Journal of Zoology, London 164*, 373-418.

Pringle, J. A. (1954). The cranial development of certain South African snakes and the relationships of these groups. *Proceedings of the Zoological Society of London 123*, 813-865.

Quattrini, D. (1952a). Richerche anatomiche e sperementali sulla autotomia della coda lucertole *(Lacerta sicula). Archo zool. ital. 37*, 131-170.

Quattrini, D. (1952b). Richerche anatomiche e sperementali sulla autotomia della coda lucertole *(Lacerta sicula). Archo zool. ital. 37*, 465-515.

Quattrini, D. (1953). Autotomia e struttura anatomica della coda dei Rettili *(Lacerta vivipara; L. viridis). Monitore zool. ital. 61*, 36-48.

Quattrini, D. (1954). Piano di autotomia e rigenerazione della coda dei Sauri *(Lacerta sicula). Archo ital. Anat. Embriol. 59*, 225-282.

Rabb, G. B. and Snedigar, R. (1960). Notes on feeding behavior of an African egg-eating snake. *Copeia* 59-60.

Radovanovic. M. (1956). Rassenbildung bei den Eidechsen auf adriatischen Inseln. *Oster. Akad. Wissensch. Mathem.-Naturwiss. Klasse, Denkschriften 110*, 1-82.

Rage, J. C. (1972). *Eryx* Daudin et *Gongylophis* Wagler (Serpentes, Boidae). Etude osteologique *Bull. Mus. natn. Hist. nat. Paris (ser. 3) 78*, 89-398.

Rai, M. (1969). Sur deux serpents recoltes en Iran *Hydrophis spiralis* Gray et *Hydrophis ornatus* Gunther. *Act. Soc. Linn. Bordeaux 106*, unpaginated.

Rathor, M. S. (1970). Movements, homing behavior and territory of the Indian sand lizard *Ophiomorus tridactylus* (Blyth). *Japanese Journal of Ecology 20*, 208-210.

Raynaud, A. (1959). Observations preliminaires sur le developpement des organes sexuels de l'embryon d'orvet *(Anguis fragilis* L.). *Bull. Soc. zool France 84*, 458-471.

Raynaud, A. (1960). Sur la differentiation sexuelle des embryons d'orvet *(Anguis fragilis* L.). *Bull. Soc. zool. France 85*, 210-230.

Raynaud, A. (1961). Le developpement des canaux de Muller chez l'embryon d'orvet *(Anguis fragilis* L.). *C. R. Soc. Biol. Paris 155*, 1893-1895.

Raynaud, A. (1962). Les premiere stades de la formation des canaux de Muller chez l'embryon d'orvet *(Anguis fragilis* L.). *Bull. biol. Fr. Belg. 96*, 281-304.

Raynaud, A. and Adrian, M. (1976). Lesions cutanees a structure papillomateuse associees a des virus chez le lezard vert *(Lacerta viridis* Laur.). *C. R. Acad. Sci. Paris D283*, 845-847.

Raynaud, A. and Adrian, M. (1977). Papillomes cutanes chez *Lacerta viridis:* etude histologique et mise en evidence de virus au moyen de la microscopie electronique. *Bull. Soc. zool. France 102*, 493-494.

Reed, C. A. and Marx, H. (1959). A herpetological collection from northeastern Iraq. *Transactions of the Kansas Academy of Sciences 62*, 91-122.

Regamey, J. (1936). Les caracteres sexuelles du lezard *(Lacerta agilis* L.). *Revue suisse zoologique 42*, 87-168.

Reid, H. A. (1976). Adder bites in Britain. *British Medical Journal 2*, 153-156.

Reid, H. A. (1977). Prolonged defibrination syndrome after bite by the carpet viper, *Echis carinatus. British Medical Journal 2*, 1326.

Richter, H. (1933). Das Zungenbein und seine Muskulatur bei den Lacertilia vera. *Jena Z. Natur. 66*, 395-480.

Richter, J. and Vozenilek, P. (1977). Imunoelektroforesa a imunoprecipitace toxinu *Vipera ammodytes. Fauna bohemiae sept.* 75-82.

Riney, T. (1953). Notes on the Syrian lizard *Acanthodactylus tristrami orientalis. Copeia* 66-67.

Roberts, J. S. and Schmidt-Nielsen, K. (1966). Renal ultrastructure and excretion of salt and water by three terrestrial lizards. *American Journal of Physiology 211*, 476-486.

Robinson, P. L. (1976). How *Sphenodon* and *Uromastyx* grow their teeth and use them. In, Bellairs, A. d'A. and Cox, C. B. (editors), *Morphology and Biology of Reptiles*, 43-64. Academic Press.

Rollinat, R. (1934). *La Vie des Reptiles de la France Centrale*. Librairie Delagrave, Paris.

Rotenberg, D., Bamberger, E. S. and Kochva, E. (1971). Studies on ribonucleic acid synthesis in the venom glands of *Vipera palaestinae* (Ophidia, Reptilia). *Biochemistry Journal 121*, 609-612.

Ruiz, M. B. (1978). Situacion actual del camaleon comun, *Chamaeleo chamaeleon* L. en la provincia de Cadiz, Espana. *Bol. Est. Centr. Ecologia 7*, 87-90.

Saint Girons, H. (1957). Le cycle sexuel chez *Vipera aspis* (L.) dans l'Ouest de la France. *Bull. Biol. France et Belge 91*, 284-350.

Saint Girons, H. (1959). Remarques histologiques sur l'hypophyse de *Vipera aspis*. *C. R. Soc. Biol. Paris 153*, 5-7.

Saint Girons, H. (1976). Les differents types de cycles sexuels des males chez les Viperes europeennes. *C. R. Acad. Sc. Paris 282*, 1017-1019.

Saint Girons, H. (1977a). Caryotypes et evolution des Viperes europeennes (Reptilia, Viperidae). *Bull. Soc. zool. France 102*, 39-49.

Saint Girons, H. (1977b). Systematique de *Vipera latastei latastei* Bosca, 1878 et description de *Vipera latastei gaditana* subsp. n. (Reptilia, Viperidae). *Revue suisse Zool. 84*, 599-607.

Saint Girons, H. (1978). Morphologie externe comparee et systematique des Viperes d'Europe (Reptilia, Viperidae). *Revue suisse Zool. 85*, 565-595.

Saint Girons, H. and Detrait, J. (1978). Communautes antigeniques des venins et systematique des Viperes europeennes. Etude immunoelectrophoretique. *Bulletin de la Societe Zoologique de France 103*, 155-166.

Saint Girons, H. and Duguy, R. (1962). Donnees histophysiologiques sur le cycle annuel de l'hypophyse chez *Vipera aspis* (L.). *Z. Zellforsch 56*, 819-853.

Saint Girons, H. and Duguy, R. (1976). Ecologie et position systematique de *Vipera seoanei* Lataste, 1879. *Bulletin de la Societe Zoologique de France 101*, 325-339.

Saint Girons, H. and Kramer, E. (1963). Le cycle sexuel chez *Vipera berus* (L.) en montagne. *Revue suisse de zoologie 70*, 191-221.

Saint Girons, H. and Saint Girons, M. C. (1956). Cycle d'activite et thermoregulation chez les reptiles (Lezards et serpents). *Vie et Milieu 7*, 133-226.

Saint Girons, M. C. (1976). Relations interspecifiques et cycle d'activite chez *Lacerta viridis* et *Lacerta agilis* (Sauria, Lacertidae). *Vie et Milieu C26*, 115-131.

Saint Girons, M. C. (1977). Le cycle d'activite chez *Lacerta viridis* et ses rapports avec la structure sociale. *Terre Vie 31*, 101-116.

Salvador, A. (1980). Interaction between the Balearic lizard *(Podarcis lilfordi)* and Eleonora's Falcon *(Falco eleonorae)*. *Journal of Herpetology 14*, 101.

Schieck, A., Kornalik, F. and Habermann, E. (1972). The prothrombin-activating principle from *Echis carinatus* venom. I. Preparation and biochemical properties. *Naunyn-Schmiedeberg Arch. exp. Path. Pharmak. 272*, 402.

Schleich, H. H. (1977). Distributional maps of reptiles of Iran. *Herpetological Review 8*, 126-129.

Schmidt, K. P. (1930). Reptiles of Marshall Field North Arabian Desert Expeditions. *Field Museum of Natural History, Zoology Series 17*, 223-230.

Schmidt, K. P. (1933). A new snake *(Rhynchocalamus arabicus)* from Arabia. *Field Museum of Natural History, Zoology Series 20*, 9-10.

Schmidt, K. P. (1939). Reptiles and amphibians from Southwestern Asia. *Field Museum of Natural History, Zoology Series 24*, 49-92.

Schmidt, K. P. (1941). Reptiles and amphibians from Central Arabia. *Field Museum of Natural History, Zoology Series 24*, 161-165.

Schmidt, K. P. (1952). Diagnoses of new amphibians and reptiles from Iran. *Natural History Miscellanea, Chicago Academy of Sciences 93,* 1-2.

Schmidt, K. P. (1953). Amphibians and reptiles of Yemen. *Fieldiana Zoology 34,* 253-261.

Schmidt, K. P. (1955). Amphibians and reptiles from Iran. *Vidensk. Medd. Dansk Naturb. Foren. Kopenhagen 117,* 193-207.

Schmidt, K. P. and Inger, R. F. (1957). *Living Reptiles of the World.* Doubleday, N.Y. 287 p.

Schmidt, K. P. and Marx, H. (1956). The herpetology of Sinai. *Fieldiana Zoology 39,* 21-40.

Schmidtler, J. J. and Schmidtler, J. F. (1970). Ein nachtgeist aus Luristan. *Aquarien Magazin 6,* 239-241.

Schmidtler, J. J. and Schmidtler, J. F. (1972). Zwerggeckos aus dem Zagros-Gebirge (Iran). *Salamandra 8,* 59-66.

Schmidt-Nielsen, K., Borut, A., Lee, P. and Crawford, E. (1963). Nasal salt excretion and the possible function of the cloaca in water conservation. *Science 142,* 1300-1301.

Schnabel, R. (1954). Papillome an einer Smaragdeidechse *(Lacerta viridis). Zoologische Garten 20,* 270-278.

Schnabel, R. (1955). Uber den Schlupfvorgang und die Resorption des Eizahns von *Natrix natrix. Z. mikrosk. anat. Forsch 61,* 487-511.

Schnabel, R. (1956). Beitrag uber die fruhen Entwicklungsstadien des Eizahns und das Abortivgebiss des Zwischenkiefers bei *Natrix natrix. Z. mikrosk. anat. Forsch. 62,* 40-50.

Schnabel, R. and Herschel, K. (1955). Uber die Entwicklung des Eizahns von *Natrix natrix. Z. mikrosk. anat. Forsch. 61,* 246-280.

Schneider, B. (1980). Eine melanistische Schlanknatter, *Coluber najadum kalymnensis* n. subsp. (Colubridae, Serpentes), von der Insel Kalymnos (Dodekanes, Agais). *Bonner zool. Beitr. 30,* 380-384.

Schulte, R. (1974). Die gefleckte Ringelnatter *(Natrix natrix persa)* vom Skutari-See. *Aquarien Terrarien -Z. 27,* 427-429.

Schulz, E. (1972). Vergleichende Beobachtungen zur Haltung und Ethologie von *Lacerta agilis agilis* und *Lacerta sicula campestris. Aquarien Terrarien 9,* 11-115.

Schweizer, H. (1921). Paarung und Fortpflanzung von *Vipera aspis* L. in Terrarium. *Aquarien Terrarien 18,* 361-363.

Scortecci, G. (1932). Rettili dello Yemen. *Atti della Societa Italiana di Scienze Naturali e del Museo Guico di Storia Naturale in Milano 71,* 39-49.

Scortecci, G. (1935). Il genre *Pristurus* nella Somalia italiana. *Atti della Societa Italiana di Scienze Naturali e del Museo Guico di Storia Naturale in Milano 74,* 118-156.

Seshadri, C. (1956). Urinary excretion in the Indian house lizard, *Hemidactylus flaviviridis* (Ruppell). *Journal of the Zoological Society of India 8,* 63-78.

Shaham, N. and Kochva, E. (1969). Localization of venom antigens in the venom gland of *Vipera palaestinae* using a fluorescent antibody technique. *Toxicon 6,* 263-268.

Shcherbak, N. N. (1971). Taxonomy of the genus *Eremias* (sauria, Reptilia) in connection with the focuses of the desert-steppe fauna development in Paleoarctic. *Vest. Zool. 5,* 48-55.

Shcherbak, N. N. (1974). *The Palearctic Desert Lizards.* Institute of Zoology, Kiev 296 p.

Shcherbak, N. N. and Golubev, M. L. (1977). Systematics of the Palearctic Geckos (Genera *Gymnodactylus, Bunopus, Alsophylax*). *Proceedings of the Zoological Institute of the Academy of Sciences, U.S.S.R. 74,* 120-133 (in Russian).

Sheppard, L. and Bellairs, A. d'A. (1972). The mechanisms of autotomy in *Lacerta*. *British Journal of Herpetology 4*, 276-286.

Shipman, W. H. and Pickwell, G. V. (1973). Venom of the yellow bellied sea snake *(Pelamis platurus)*. Some physical and chemical properties. *Toxicon 11*, 375-377.

Simms, C. (1972). Shift in a population of northern vipers. *British Journal of Herpetology 4*, 268-271.

Simms, C. (1976). The sand lizard in north-east England. *British Journal of Herpetology 5*, 522-525.

Singh, L., Purdom, I. F. and Jones, K. W. (1976). The chromosomal localisation of satellite DNA in *Ptyas mucosus* (Ophidia, Colubridae). *Chromosoma (Berlin) 57*, 177-184.

Sjongren, S. J. (1945). Uber die Embryona lentwicklung des Sauropsidenmagens. *Acta anat. Suppl. 2*, 1-223.

Skoczylas, R. (1970a). Influence of temperature on gastric digestion in the grass snake, *Natrix natrix* L. *Comparative Biochemistry and Physiology 33*, 793-804.

Skoczylas, R. (1970b). Salivary and gastric juice secretion in the grass snake, *Natrix natrix* L. *Comparative Biochemistry and Physiology 35*, 885-903.

Smirnov, S. V. (1979). Structure and function of acoustic analyzer in *Eumeces schneideri* Daudin (Squamata, Scincidae). *Zool. Zhur. 58*, 1425-1428 (in Russian).

Smith, M. A. (1926). *Monograph of the sea-snakes (Hydrophiidae)*. B.M.N.H. 1-130.

Smith, M. A. (1973). *British Amphibians and Reptiles*. Collins, London (edition 5) 322 p.

Sokolov, V. E. (1966). Water content in tissues of some desert animals. *Zool. Zhur. 45*, 776-777 (in Russian).

Somani, P. and Arora, R. B. (1962). Mechanism of increased capillary permeability induced by *Echis carinatus* (saw-scaled viper) venom: A possible new approach to the treatment of viperine snake poisining. *J. Pharm. Pharmac. 14*, 394-395.

Spellerberg, I. F. (1974). Influence of photoperiod and light intensity on lizard voluntary temperatures. *British Journal of Herpetology 5*, 412-420.

Spellerberg, I. F. (1976). Adaptations of reptiles to cold. In, Bellairs, A. d'A. and Cox, C. B. (editors), *Morphology and Biology of Reptiles*, 261-285. Academic Press.

Spellerberg, I. F. (1977). Behaviour of a young smooth snake, *Coronella austriaca* Laurenti. *Biological Journal of the Linnean Society 9*, 323-330.

Spellerberg, I. F. and Phelps, T. (1975). Preliminary investigations into the voluntary temperatures of the smooth snake, *Coronella austriaca*. *Copeia* 183-185.

Spellerberg, I. F. and Phelps, T. (1977). Biology, general ecology and behaviour of the snake, *Coronella austriaca* Laurenti. *Biological Journal of the Linnean Society 9*, 133-164.

Spellerberg, I. F. and Smith, N. D. (1975). Inter- and intra-individual variation in lizard voluntary temperatures. *British Journal of Herpetology 5*, 496-504.

Spitz, F. (1971). Quelques donnes sur les lezards (*Lacerta viridis* et *L. agilis*) marques a la point de Arcay (Vendee). *Terre et la Vie 25*, 86-95.

Stemmler, O. (1958). *Eryxjaculus turcicus* Olivier — die europaische Riesenschlange. *Z. Vivaristik 4*, 89-100.

Stemmler, O. (1967). Der Kommentkampf von *Vipera lebetina schweizeri*. *Aqua Terra 4*, 89-91.

Steward, J. W. (1958). The dice snake *(Natrix tessellata)* in captivity. *British Journal of Herpetology 2*, 122-126.

Steward, J. W. (1965). Territorial behaviour in the wall lizard *Lacerta muralis*. *British Journal of Herpetology 3*, 224-229.

Steward, J. W. (1971). *The Snakes of Europe*. David and Charles. 238 p.

Stimson, A. F. (1969). Liste der rezenten Amphibien und Reptilien. Boidae (Boinae + Bolyeriinae + Loxoceminae + Pythoninae). *Das Tierreich 89*, 1-49.

Stolk, R. (1953). Hyperkeratosis and carcinoma planocellulare in the lizard, *Lacerta agilis*. *Koninklijke Nederlandash Akademie von Wetenschappen C56*, 157-163.

Street, D. J. (1973). Notes on the reproduction of the Southern smooth snake *(Coronella girondica)*. *British Journal of Herpetology 4*, 335-337.

Subba Rao, M. V. (1975a). Influence of body weight, sex and temperature on heart beat in the garden lizard. *Calotes versicolor. British Journal of Herpetology 5*, 464-466.

Subba Rao, M. V. (1975b). Studies on the food and feeding behaviour of the Agamid garden lizard *Calotes versicolor. British Journal of Herpetology 5*, 467-470.

Sweeney, R. C. H. (1971). *Snakes of Nyasaland, with new added corrigenda and addenda*. Asher, Amsterdam, 200 p.

Swiezawska, K. (1949). Colour discrimination of the sand lizard *Lacerta agilis* L. *Bull. Int. Acad. Pol. Sci. Let. (ser. B) Sci. Nat.* 1-20.

Taborska, E. (1971). Intraspecies variability of the venom of *Echis carinatus. Physiol. bohemoslov. 20*, 307-318.

Taddei, C. (1972). Ribosome arrangement during oogenesis of *Lacerta sicula* Raf. *Experimental Cell Research 70*, 285-292.

Tansley, K. (1959). The retina of two nocturnal geckos, *Hemidactylus turcicus* and *Tarentola mauritanica. Pflugers Arch. ges. Physiol. 268*, 213-220.

Taylor, E. H. (1935). A taxonomic study of the cosmopolitan scincoid lizards of the genus *Eumeces* with an account of the distribution and relationships of its species. *Kansas University Science Bulletin 23*, 1-643.

Tercafs, R. R. (1962). Observations ecologiques dans le massif du Tibesti (Tchad). *Rev. Zool. Bot. Afr. 66*, 107-126.

Tertyshnikov, M. F. (1970). Food of variegated lizard (*Eremias arguta deserti* Gmel. 1788) in central cis-Caucasia. *Soviet Journal of Ecology 1*, 344-349.

Thapar, G. S. (1921). On the venous system of the lizard *Varanus bengalensis. Proceedings of the zoological society of London* 487-492.

Theakston, R. D. G., Lloyd-Jones, M. J. and Reid, H. A. (1977). Micro-elisa for detecting and assaying snake venom and venon-antibody. *Lancet* 639-641.

Thomas, E. (1955). Der Kommentkampf der Kreuzotter (*Vipera berus* L.) *Naturwissensch. 42*, 539.

Thomas, E. (1961). Fortpflanzungskampfe bei Sandottern *(Vipera ammodytes). Zool. Anz. Suppl. 24*, 502-505.

Thomas, E. (1969). Jungtierkampfe bei *Vipera a. ammodytes* (Serpentes, Viperidae). *Salamandra 5*, 141-142.

Thomas, E. (1971). *Vipera ammodytes montandoni* (Viperidae). Kommentkampf der Mannchen. *Publ. Wiss. Film Sekt., Biol. 4*, 178-188.

Thomas, E. (1972). *Bitis arietans* (Viperidae). Kommentkampf der Mannchen. *Publ. Wiss. Film Sekt., Biol. 5*, 291-299.

Thorpe, R. S. (1975a). Biometric analysis of incipient speciation in the Ringed Snake, *Natrix natrix* (L.). *Experientia 31*, 180-182.

Thorpe, R. S. (1975b). Quantitative handling of characters useful in snake systematics with particular reference to intraspecific variation in the Ringed Snake *Natrix natrix* (L.). *Biological Journal of the Linnean Society 7*, 27-43.

Thorpe, R. S. (1979). Multivariate analysis of the population systematics of the ringed snake *Natrix natrix* (L.). *Proceedings of the Royal Society of Edinburgh 78B*, 1-62.

Thorpe, R. S. (1980a). A comparative study of ordination techniques in numerical taxonomy in relation to racial variation in the ringed snake *Natrix natrix* (L.). *Biological Journal of the Linnean Society 13*, 7-40.

Thorpe, R. S. (1980b). Microevolution and taxonomy of European reptiles with particular reference to the grass snake *Natrix natrix* and the wall lizards *Podarcis sicula, P. melisellensis. Biological Journal of Linnean Society 14*, 215-233.

Thorpe, R. S. (1981). Racial divergence and subspecific status of the Gotland grass snake: a comment on Nilson & Andren's paper. *Zoological Journal of the Linnean Society 72*, 369-370.

Throckmorton, G. S. (1976). Oral food processing in two herbivorous lizards, *Iguana iguana* (Iguanidae) and *Uromastyx aegyptia* (Agamidae). *Journal of Morphology 148*, 363-390.

Tiedemann, F. and Haupl, M. (1978). Ein weiterer nachweis von *Elaphe quatuor-lineata sauromates* aus Syrien (Reptilia: Serpentes: Colubridae). *Salamandra 14*, 212-214.

Tornier, G. (1905). Schildkroten und Eidechsen aus Nordost-Afrika und Arabien. *Zool. Jahrb. Syst. 22*, 365-388.

Tortonese, E. (1948). Osservazioni biologiche su Anfibi e Rettili di Rodi, Anatolia, Palestina e Egitto. *Arch. Zool. Ital., Padova 33*, 377-402.

Tortonese, E. and Lanza, B. (1968). *Pesci, Anfibi et Rettili*. Aldo Martello Editore, Milan.

Trost, E. (1953). Die Entwicklung, Histogenese und Histologie der Epiphyse, der Paraphyse, der Velum transversum, des Dorsalsackes und des subcommissuralen Organs bei *Anguis fragilis, Chalcides ocellatus* und *Natrix natrix*. *Acta anat. 18*, 326-342.

Tu, A. T., Lin. T. S. and Bieber A. L. (1975). Purification and chemical characterization of the major neurotoxin from the venom of *Pelamis platurus*. *Biochemistry N.Y. 14*, 3408-3413.

Tuck, R. G. (1971). Amphibians and reptiles from Iran in the United States National Museum Collection. *Bulletin of the Maryland Herpetological Society 7*, 46-86.

Tzellarius, A. Y. (1977). A contribution to the ecology of *Eremias grammica* (Lacertidae, Sauria) in East Karakumy. *Zool. Zhur. 56*, 224-231 (in Russian).

Underwood, G. (1954). On the classification and evolution of geckos. *Proceedings of the zoological society of London 124*, 469-472.

Underwood, G. (1967). A contribution to the classification of snakes. B.M.N.H. 653, 1-179.

Underwood, G. (1971). A modern appreciation of Camp's "Classification of the lizards". In, *Camp's Classification of the lizards*. Facsimile reprint. S.S.A.R.

Uzzell, T. and Darevsky, I. S. (1973a). Electrophoretic examination of *Lacerta mixta* a possible hybrid species (Sauria, Lacertidae). *Journal of Herpetology 7*, 11-15.

Uzzell, T. and Darevsky, I. S. (1973b). The relationships of *Lacerta portschinskii* and *Lacerta raddei* (Sauria, Lacertidae). *Herpetologica 29*, 1-6.

Uzzell, T. and Darevsky, I. S. (1975). Biochemical evidence for the hybrid origin of the parthenogenetic species of the *Lacerta saxicola* complex (Sauria: Lacertidae), with a dicussion of some ecological and evolutionary implications. *Copeia* 204-222.

Vainio, J. (1931). Zur Verbreitung und Biologie der Kreuzotter, *Vipera berus* (L.) in Finland. *Ann. Zool. Soc. Vanamo 12*, 1-19.

Van der Walt, S. J. (1972). Studies on *Bitis arietans* venom. IV. Association of protease A. *Z. phys. Chem. 353*, 1217-1227.

Van der Walt, S. J. and Joubert, F. J. (1971). Studies on puff adder *(Bitis arietans)* venom. I. Purification and properties of protease A. *Toxicon 9*, 153-161.

Van der Walt, S. J. and Joubert, F. J. (1972). Studies on the venom of puff adder *(Bitis arietans)*. II. Specificity of protease A. *Toxicon 10*, 341-356.

Vashetko, E. V. (1969). A contribution to the ecology of *Eremias strauchii* in the southwest Turkemia. *Zoological Journal (Moscow) 48*, 1893-1895 (in Russian).

Vashetko, E. V. (1971). Ecology of the lizard *Eremias velox velox* in the Fergana Valley. *Zoological Journal (Moscow) 51*, 153-155 (in Russian).

Venzmer, G. (1919). Beitrage zur Kenntnis der Reptilien und Amphibien-Fauna d. cilicischen Taurus. *Sitz. Ber. Ges. nat. forsch. Freunde Berlin 7*, 209-251.

Viitanen, P. (1967). Hibernation and seasonal movements of the viper *Vipera berus berus* (L.) in southern Finland. *Ann. Zool. Fenn. 4*, 472-546.

Vladescu, C. (1965a). Glycaemia in the *Vipera berus*. *Rev. Roumaine Biol., ser. zool. 10*, 43-46.

Vladescu, C. (1965b). Researches on normal glycemia and induced hyperglycemia in *Lacerta agilis chersonensis*. *Rev. Roumaine Biol., ser. zool. 10*, 171-175.

Vladescu, C. and Motelica, I. (1965). The influence of insulin on glycemia in *Lacerta agilis chersonensis*. *Rev. Roumaine Biol., ser. zool. 10*, 451-456.

Vogel, Z. (1964). *Reptiles and Amphibians, their care and behaviour*. Studio Vista 228 p.

Volsoe, H. (1944). Structure and seasonal variation of the male reproductive organs of *Vipera berus*. *Spolia Zool. Mus. Hauniensis Copenhagen 5*, 1-172.

Voris, H. K. (1977). A phylogeny of the sea snakes (Hydrophiidae). *Fieldiana Zoology 70*, 79-166.

Walter, H. (1967). Zur Lebensweise von *Lacerta erhardii*. *Bonn Zool. Beitr. 18*, 216-220.

Warburg, M. R. (1964). Observations on the microclimate in habitats of some desert vipers in the Negev, Arava and Dead Sea regions. *Vie Milieu 15*, 1017-1041.

Warrell, D. A. and Arnett, C. (1976). The importance of bites by the saw-scaled or carpet viper *(Echis carinatus)*: epidemiological studies in Nigeria and a review of the world literature. *Acta trop. Sep. 33*, 307-341.

Warrell, D. A., Barnes, H. J. and Piburn, M. F. (1976). Neurotoxic effects of bites by the Egyptian cobra *(Naja haje)* in Nigeria. *Transactions of the Royal Society for Tropical Medicine and Hygiene 70*, 78-79.

Warrell, D. A., Davidson, N. Mc., Greenwood, B. M., Ormerod, L. D., Pope, H. M., Watkins, B. J. and Prentice, C. R. M. (1977). Poisoning by the bites of the saw-scaled or carpet viper *(Echis carinatus)* in Nigeria. *Quarterly Journal of Medicine (new ser.) 46*, 33-62.

Warrell, D. A., Davidson, N. Mc., Ormerod, L. D., Pope, H. M., Watkins, B. J., Greenwood, B. M. and Reid, H. A. (1974). Bites by the saw-scaled or carpet viper *(Echis carinatus)*: trial of two specific antivenoms. *British Medical Journal 4*, 437-440.

Weber, H. (1957). Vergleichende Untersuchung des Verhaltens von Smaragdeidechsen *(L. viridis)*, Mauereidèchsen *(L. muralis)* und Perleidechsen *(L. lepida)*. *Z. Tierpsychol. 14*, 448-472.

Weber-Semenoff, D. (1977). Mating behaviour of *Eirenis collaris* (Menetries) in captivity. *Proceedings of the Zoological Institute of the Academy of Sciences U.S.S.R. 74*, 36-38 (in Russian).

Weigmann, R. (1932). Jahrescyklische Veranderungen im Funktionszustand der Schilddruse und im Stoffumsatz von *Lacerta vivipara* Jacq. *Z. wiss. Zool. 142*, 491-509.

Welch, K. R. G. (1980). A comment on the European colubrid genus *Haemorrhois*. *South Western Herpetological Society Bulletin 3*, 14-15.

Welch, K. R. G. (1982). *Herpetology of Africa. A Checklist and Bibliography of the Orders Amphisbaenia, Sauria and Serpentes*. R. E. Krieger, 293 p.

Wermuth, H. (1965). Liste der rezenten Amphibien und Reptilien. Gekkonidae, Pygopodidae, Xantusiidae. *Das Tierreich 80*, 1-246.

Wermuth, H. (1967). Liste der rezenten Amphibien und Reptilien. Agamidae. *Das Tierreich 86*, 1-127.

Wermuth, H. (1969). Liste der rezenten Amphibien und Reptilien. Anguidae. Anniellidae, Xenosauridae. *Das Tierreich 90*, 1-42.

Werner, D. (1972). Boebachtungen an *Ptyodactylus hasselquistii guttatus* (Gekkonidae). *Verhandl. Naturf. Ges. Basel 82*, 54-87.

Werner, F. (1898). Uber einige neue Reptilien und einen neuen Frosch aus dem cilicischen Taurus. *Zool. Anz. Leipzig 21,* 217-223.

Werner, F. (1900). Beschreibung einer bisher noch unbekannten Eidechse aus Kleinasien *Lacerta anatolica. Anz. Akad. Wiss. Wien 25,* 269-271.

Werner, F. (1902). Die Reptilien und Amphibienfauna von Kleinasien. *Sitzber. Akad. Wiss. Wien, Math.-Nat. Cl. Wien 111,* 1057-1121.

Werner, F. (1903). Uber Reptilien und Batrachier aus West-Asien (Anatolien und Persien). *Zool. Jahrb. Syst., Jena 19,* 329-346.

Werner, F. (1935). Reptilien der Agaischen Inseln. *Sitzber. Akad. Wiss. Wien 144,* 81-117.

Werner, F. (1938a). Reptilien aus dem Iran und Belutschistan. *Zool. Anz. 121,* 265-271.

Werner, F. (1938b). Eine verkannte Viper (*Vipera palaestinae* n. sp.). *Zool. Anz. 122,* 313-318.

Werner, F. (1938c). Die Amphibien und Reptilien Griechenlands. *Zoologica 94,* 1-117.

Werner, F. (1939). Die Amphibien und Reptilien von Syrien. *Abh. ber. Mus. Nat.-U. Heimatk. (Naturk. Vorgesch.) Magdeburg 7,* 211-223.

Werner, Y. L. (1964). Frequences of regenerated tails, and structure of caudal vertebrae in Israeli desert geckos (Reptilia: Gekkonidae). *Israel Journal of Zoology 13,* 134-136.

Werner, Y. L. (1965). Ueber die israelischen Geckos der Gattung *Ptyodactylus* und ihre Biologie. *Salamandra 1,* 15-25.

Werner, Y. L. (1968). Distribution of the Saharan *Sphenops sepsoides* (Reptilia: Scincidae) in Israel and Jordan. *Herpetologica 24,* 238-242.

Werner, Y. L. (1971). Lizards and snakes from Transjordan, recently acquired by the British Museum (Natural History). *Bulletin of the British Museum (Natural History), Zoology 21,* 213-256.

Werner, Y. L. and Drook, K. (1967). The multipartite testis of the snake *Leptotyphlops phillipsi. Copeia* 159-163.

Werner, Y. L. and Goldblatt, A. (1978). Body temperature in a basking Gekkonid lizard, *Ptyodactylus hasselquistii* (Reptilia Lacertilia Gekkonidae). *Journal of Herpetology 12,* 408-411.

Wettstein, O. (1928). Amphibien und Reptilien aus Palestina und Syrien. *Sitzber. Akad. Wiss. Wien Abt. 1, 137,* 773-785.

Wettstein, O. (1951). Ergebnisse der osterreichischen Iran-Expedition 1949/50. *Sitzber. Akad. Wiss. Wien Math.-Nat. Kl. Wien 160,* 427-448.

Wettstein, O. (1953). Herpetologia aegaea. *Sitzber. Akad. Wiss. Wien Math.-Nat. Kl. Wien 162,* 651-833.

Wettstein, O. (1957). Nachtrag zu meiner Herpetologia aegaea. *Sitzber. Akad. Wiss. Wien Math.-Nat. Kl. Wien 166,* 123-164.

Wettstein, O. (1960a). Contribution a l'etude de la faune d'Afghanistan. 3. Lacertilia aus Afghanistan. *Zoologischer Anzeiger 165,* 58-63.

Wettstein, O. (1960b). Drei seltene Eidechsen aus Sudwest-Asien. *Zoologischer Anzeiger 165,* 190-193.

Wettstein, O. (1964). Herpetologisch neues aus Rhodos. *Senckenberg. biol. 45,* 501-504.

Wettstein, O. (1967). Ergebnisse zoologischer Sammelreisen in der Turkei: Versuch einer Klarung des Rassenkreises von *Lacerta danfordi* GTHR 1876. *Annls. naturh. Mus. Wien 70,* 345-356.

Willemse, G. T., Hattingh, J. and Coetzee, N. (1979). Precipitation of human blood clotting factors by puff-adder *(Bitis arietans)* venom. *Toxicon 17,* 331-335.

Willemse, G. T., Hattingh, J., Karlsson, R. M., Levy, S. and Parker, C. (1979). Changes in composition and protein concentration of puff-adder *(Bitis arietans)* venom due to frequent milking. *Toxicon 17*, 37-42.

Witte, G. F. de (1948). *Faune de Belgique. Amphibiens et Reptiles* (2nd edition). Musee Royal d'Histoire Naturelle de Belgique, Bruxelles.

Witte, G. F. de (1973). Description d'un Gekkonidae nouveau de l'Iran. *Bull. Inst. v. Sci. Nat. Belg. 49*, 1-6.

Witte, G. F. de and Laurent, R. (1947). Revision d'un groupe de Colubridae Africains. Genres *Calamelaps, Miodon, Aparallactus* et formes affines. *Memoires de Musee Royal d'Histoire Naturelle de Belgique (ser. 2) 29*, 1-134.

Wittig, W. (1976). Zucht und Haltung der Pityusen-Eidechse. *Aquar. Terr. Berlin 23*, 388-389.

Wolter, M. (1924). Die Giftdruse von *Vipera berus. Jena Z. Naturw. 60*, 305-362.

Wood, S. F. (1938). Variations in the cytology of the blood of geckos, *Tarentola mauritanica*, infected with *Haemogregarina platydactyli, Trypanosoma platydactyli* and *Pirhemocyton tarentolae. University of California Publications, Zoology 41*, 9-21.

Zavattari, E. (1929). Anfibi e rettili. Ricerche faunistiche nelle Isole Italiane dell' Egeo. *Arch. Zool. Ital., Padova 13*, 31-36.

Zeiller, W. (1969). Maintenance of the yellow bellied sea snake *Pelamis platurus* in captivity. *Copeia* 407-408.

Zimmerman, H., Habermann, E. and Lasch, H. G. (1971). Der Einflu von Gift der Sandrasselotter *(Echis carinatus)* auf die Hamostase. *Thromb. Diath. Haemorrh. 25*, 425-437.

Zingg. A. (1968). Die Kykladenviper, *Vipera lebetina schweizeri* Werner 1935. *Aquaterra 5*, 73-76.

Zinner, H. (1967). Herpetological collection trips to the Lebanon in 1965 and 1966. *Israel Journal of Zoology 16*, 49-58.

Zinner, H. (1977). The status of *Telescopus hoogstraali* Schmidt and Marx 1956 and the *Telescopus fallax* Fleischmann 1831 complex (Reptilia, Serpentes, Colubridae). *Journal of Herpetology 11*, 207-212.

Zwinenberg, A. J. (1979). Biologie en status van de Levantijnse Adder van de Cycladen, *Vipera lebetina schweizeri. Lacerta 9*, 138-146.

INDEX TO GENERA

127

INDEX TO SPECIES AND SUBSPECIES